# Imported Virus Infections

Edited by
Tino F. Schwarz
Günter Siegl

Archives of Virology
Supplement 11

SpringerWienNewYork

Dr. Tino F. Schwarz
Stiftung Juliusspital
Würzburg, Federal Republic of Germany

Prof. Dr. Günter Siegl
Institute for Clinical Microbiology
and Immunology,
St. Gallen, Switzerland

© 1996 Springer-Verlag/Wien


Typesetting: Thomson Press (India) Ltd., New Delhi


Graphic design: Ecke Bonk

Printed on acid-free and chlorine-free bleached paper

With 68 Figures

**Library of Congress Cataloging-in-Publication Data**

Imported virus infections/Tino F. Schwarz and Günter Siegl (eds.).
    p.    cm. − − (Archives of Virology. Supplement, ISSN 0939-1983; 11)
    Based on papers presented at an international conference on imported virus infections,
held at the Max von Pettenkofer Institute, University of Munich, Germany on March 31
to April 1, 1995.
    Includes bibliographical references.
    ISBN-13: 978-3-211-82829-8         e-ISBN-13: 978-3-7091-7482-1
    DOI: 10.1007/978-3-7091-7482-1
    1. Virus diseases − − Epidemiology − − Congresses.     I. Schwarz, Tino F.
II. Siegl, Günter, 1940−     .     III. Series: Archives of Virology.
Supplement; 11.
    [DNLM: 1. Virus Diseases − − transmission − − congresses. 2. Virus
Diseases − − epidemiology − − congresses.    3. Virus Diseases − − therapy − −
congresses.    4. Travel − − congresses.    W1 AR49LA v. 11 1996 / WC 500
I34 1996]
RA644. V55I47        1996
616′.0194 − − dc20
DNLM/DLC
for Library of Congress                                                                    96-8183
                                                                                                       CIP


ISSN 0939-1983

# Preface

Mobility of large parts of the human population, whether related to commercial necessity, touristic activities or to migration induced by war and social pressure, carried and carries the risk of spreading infections. Modern air travel effectively circumvents existing quarantine regulations as infected individuals thereby can reach almost every geographic location while still in the incubation phase of the disease. Hence, infections previously restricted to distinct regions due to their strict association with non-human reservoirs or vectors can suddenly surface in non-endemic areas where lacking experience and technical means make clinical and laboratory diagnosis difficult. Excellent examples for such situations are many vector- or rodent-borne viruses but also hepatitis viruses, the human immunodeficiency virus and, last but not least, filoviruses.

The following articles are based on papers presented at an international symposium on "Imported Virus Infections" held at the Max von Pettenkofer Institute, University of Munich, Munich, Germany on March 31 to April 1, 1995. They illustrate today's knowledge on the epidemiology, dynamics of spread, as well as the frequently limited possibilities of prevention and therapeutic treatment of associated disease. Special emphasis was placed on filovirus infections which, as if to highlight the topics of the symposium, reappeared and spread in Zaire in the first half of 1995.

The symposium was dedicated to the memory of Friedrich Deinhardt M.D., virologist, professor and director of the Max von Pettenkofer Institute for Hygiene and Medical Microbiology from 1977 until he died on April 30, 1992. Throughout his scientific career he had a continued interest in emerging virus diseases and till his death actively supported research programs on human immunodeficiency virus and hepatitis C virus infections at the institute. He certainly would have appreciated this meeting of experts as well as the lively discussions concerning the various topics at the symposium.

The symposium was sponsored by contributions from pharmaceutical companies and private agencies, which we hereby gratefully acknowledge:
Abbott GmbH, Germany, Baxter Deutschland GmbH, Germany, Behringwerke AG, Germany, Biotest Pharma GmbH, Germany, Boehringer Mannheim GmbH, Germany, Deutsche Vereinigung zur Bekämpfung der Viruskrankheiten e.V., Germany, Deutsche Wellcome GmbH, Germany, Dr. Karl Thomae GmbH, Germany, Förderverein des Deutschen Grünen Kreuzes, Germany, Immuno GmbH, Germany, Institut Virion GmbH, Germany, Pasteur-Merieux-MSD GmbH, Germany, Procter & Gamble GmbH, Germany, Sanofi GmbH, Germany, SmithKline Beecham GmbH, Germany.

*Tino F. Schwarz*
*Günter Siegl*

# Contents

## Travel-related vector-borne virus infections and emerging virus diseases

## Filovirus infections

## Hepatitis viruses and human immunodeficiency virus

Contents

# Travel-related vector-borne virus infections
## and emerging virus diseases

Arch Virol (1996) [Suppl] 11: 3–11

# Imported vector- and rodent-borne virus infections – an introduction

**T. F. Schwarz**

Max von Pettenkofer Institute for Hygiene and Medical Microbiology, Ludwig
Maximilians University, Munich, Federal Republic of Germany

**Summary.** Travel is a potent force in the emergence of virus infections. Migration of humans and animals has been the pathway for disseminating virus diseases throughout history. In recent years, dengue virus has been identified as the most important travel-related, vector-borne virus disease. Other vector-borne virus infections, such as sandfly fever, Rift Valley fever, chikungunya fever and Japanese encephalitis, have been diagnosed in travelers returning from endemic areas. Crimean-Congo haemorrhagic fever may not only be imported by infected live stock, but also by travelers. Of rodent-borne virus infections, Lassa fever has been diagnosed occasionally in travelers returning from endemic areas. The potential impact of imported filoviruses is currently discussed.

## Introduction

Importation of virus infections is equivalent to the transport of an agent from an endemic to a non-endemic area. In recent years, most studies of imported virus infections in Europe dealt with hepatitis A virus, hepatitis B virus, and human immuno-deficiency virus, and imported vector- and rodent-borne virus infections received little attention. Although imported arbovirus infections and a few cases of rodent-borne infections were reported, the true incidence of cases in non-endemic countries was unknown. However, the gravity of this problem is illustrated by the diagnosis of the following vector-borne virus infections which were imported from endemic countries to the Netherlands from 1977 to 1980: Bwamba virus (Cameroon), chikungunya virus (Cameroon, Niger, Nigeria, Zaire), Chagres virus (Costa Rica), Colorado tick fever virus (USA), Mayaro virus (Surinam), Rift Valley fever virus (Tanzania), Ross River virus (Samoa), and dengue virus (Thailand, Indonesia, Surinam, Curaçao) [40].

Several factors may underly the increase in the incidence of imported vector- and rodent-borne virus infections. Travelers returning from endemic tropical and subtropical countries probably account for most cases of imported virus infection; according to the World Tourism Organization, Madrid, Spain, 528 million people vacationed in foreign countries in 1994 with Germans and Americans accounting for 65 and 47 million visits abroad, respectively. Other possible sources for the importation of virus infections are transit passengers, asylum-seekers and refugees from endemic areas in Third World countries. Similarly, aid workers and military staff, e.g. United Nations troops, stationed in endemic countries may be exposed to infectious agents, and can import the diseases upon return to their home countries. Also, virus infections have been introduced by animals imported from endemic areas.

## Current frequency of imported vector- and rodent-borne virus infections

Several studies and case reports have demonstrated the importation of vector- and rodent-borne virus infections in recent years.

### *Dengue fever*

Of all the vector-borne infections, dengue fever appears to be the most frequently imported virus disease from endemic to non-endemic countries. During the first half of 1995, approximately 250 clinical cases of imported dengue fever were diagnosed in Germany (J. L'age-Stehr, Berlin pers. comm.). Dengue viruses are endemic and epidemic in tropical and subtropical regions, especially in the Caribbean basin, South America, Central America, some African countries, islands of the Indian Ocean, southeast Asia, Australia, and Oceania; they are transmitted by mosquitos (*Aedes*) [15]. In some endemic areas, such as the Americas, and southeast Asia, the incidence of this infection has been increasing dramatically in recent years [15]. Dengue infection may be asymptomatic, or overt disease may range from a mild, febrile illness (classical dengue fever) to severe and fatal haemorrhagic disease (dengue haemorrhagic fever, dengue shock syndrome). Most infections were reported in travelers returning from endemic areas in southeast Asia, especially Thailand, the Caribbean Islands and Oceania [3, 4, 8, 17, 21, 25, 29, 32, 36, 42, 43] or in military personnel stationed in the Philippines [18], Haiti [5] or Somalia [23]. In most patients, dengue fever was diagnosed whereas haemorrhagic dengue fever or dengue shock syndrome still seem to be rare complications in travelers from non-endemic countries [6, 8, 25, 29]. With the growing spread of dengue viruses and their vectors, *Aedes aegypti* and *Aedes albopictus*, it is likely that imported cases will increase but as this disease is not notifiable, and diagnostic tests are rarely performed in non-endemic countries, the true extent of the problem is not known.

### *Sandfly fever*

Sandfly fever viruses belong to the *Phlebovirus* genus of the *Bunyaviridae*. Three serotypes, Sicilian (SFSV), Naples (SFNV) and Toscana (TOSV) are medically significant because they cause sandfly fever (pappataci fever). This infection gained some importance in the Mediterranean region during World War II, afflicting 19 000 soldiers requiring hospitalization. SFSV and SFNV cause a febrile illness with headaches and myalgia lasting for several days whereas TOSV is more virulent, causing high fever, severe headaches, aseptic meningitis and meningoencephalitis. Convalescence may be prolonged. Sandfly fever is becoming more significant as a travel-related disease, and several cases of sandfly fever in travelers and soldiers infected during visits to Italy, Cyprus, Spain, Tunisia, and Central Asia have been reported recently [2, 10–12, 14, 33–35]. The reduced use of pesticides in the Mediterranean region has led to an increase in the number and density of sandflies. As these insects are the vector for sandfly fever and leishmaniasis, more clinical cases may be expected in endogenous populations and travelers in the future.

## Rift Valley fever

Rift Valley fever virus is widely distributed throughout sub-Saharan Africa. So far, the virus has been isolated in Burkina Faso, the Central African Republic, Egypt, Guinea, Kenya, Madagascar, Mauritania, Mozambique, Namibia, Nigeria, Senegal, South Africa, Sudan, Tanzania, Uganda, Zaire, Zambia, and Zimbabwe. In the past 20 years, epizootics affecting livestock and humans occurred in Egypt, Mauritania, and Senegal. The latest epidemic started in Sudan and Egypt in 1993, and lasted until 1994 [1]. In the late 1970s, imported cases of Rift Valley fever acquired in Egypt were reported in travelers returning to the United Kingdom [9, 45], and in Swedish UN soldiers stationed in Egypt and in the Sinai [31]. Since this disease occurs in irregular cycles in Africa, infections may be expected only during epizootic outbreaks.

## Lassa fever

Lassa fever virus is endemic in West Africa ranging from Guinea to Nigeria. In the last 15 years, several cases of imported Lassa fever were observed in the USA [20, 49], Canada [27], the United Kingdom [7], and Japan [19]. Nosocomial spread was not observed with the most recent cases. With further political and economic destabilisation of some West African countries, established programmes to control Lassa fever virus, e.g. in Sierra Leone, may fail, subsequently leading to locally increased endemicity and an increased likelihood of future importation in travelers or endogenous persons returning from West Africa.

## Chikungunya fever

Chikungunya virus is endemic in most parts of Africa and Southeast Asia, and epidemics have occurred in sub-Saharan Africa, India, Thailand, and the Philippines. Until recently, imported infections with chikungunya virus in travelers were rarely reported [16, 40]. Chikungunya fever is a dengue-like illness, and in areas where dengue is endemic, clinical diagnosis of chikungunya fever may be difficult, requiring laboratory confirmation. Since the virus is present in Africa and Asia, imported infections may be expected in the future.

## Japanese encephalitis

Japanese encephalitis virus (JEV) is endemic in tropical areas of south and southeastern Asia [41]. Recently, an outbreak of JEV occurred among residents of an island of Australia's Torres Strait, which lies between mainland Queensland and Papua New Guinea [47]. Infections with JEV acquired in Southeast Asia, were reported in travelers returning from Bali, Indonesia [28, 44]. Although control measurements have been effectively implemented in some East Asian countries, JEV circulates in countries such as Thailand, Vietnam, Myanamar, Nepal, Bangladesh, and Sri Lanka [41]. As an active vaccine is widely available for travelers, infection can be prevented.

### Crimean-Congo haemorrhagic fever

Crimean-Congo haemorrhagic fever virus (C-CHFV) is widely distributed throughout the arid regions of Africa, the Middle East, south and eastern Europe and Asia. C-CHFV may cause outbreaks or isolated cases among farm and abattoir workers, and veterinary surgeons. The infection is enzootic, but mainly asymptomatic in many species such as cattle, sheep, goats, camels and hares. Humans become infected by bites of *Hyalomma* ticks or through close contact with infected animals or humans. Nosocomial outbreaks of C-CHFV infections have been reported repeatedly [13, 38, 39]. Mortality rates ranged from 10 to 80% in different outbreaks. C-CHFV may be introduced in a non-endemic area through the import of livestock from endemic areas, and thus cause outbreaks. The recent C-CHFV outbreak in the United Arab Emirates was thought to have been caused by the importation of infected sheep from neighbouring countries (pers. comm., H. Nsanze, Al Ain). Travelers may be infected during visits of endemic areas.

### Filovirus-associated haemorrhagic fever

Since the first outbreak of Marburg virus in the homonymous city in Germany in 1967, associated with imported African green monkeys (*Cercopithecus aethiops*) from Uganda [37], sporadic cases of Marburg virus infections have been reported in travelers infected in Zimbabwe/South Africa in 1975, and Kenya in 1980 and in 1987 [39]. Recently, the first cases of suspected imported filovirus haemorrhagic fever were reported in Sweden [24] and in Switzerland [26]. In 1989, a simian haemorrhagic fever virus, now designated Ebola Reston, imported to a primate center in Reston, VA, USA, from the Philippines, only caused asymptomatic infections in staff working with infected monkeys [22]. In 1992, a filovirus, identical to Ebola Reston, was isolated from cynomolgus monkeys imported from the Philippines into Italy; it is noteworthy that these animals were obtained from the same exporter in the Philippines [46]. These two incidents demonstrated the need for better safety standards, particularly those concerning infectious diseases, for animal imports, especially primates. In 1994, Ebola virus was isolated from a Swiss ethologist who got infected during autopsies of chimpanzees in Ivory Coast [26]. This Ebola strain, designated Ebola Côte d'Ivoire, had caused an outbreak of haemorrhagic fever in chimpanzees. The year of 1995 witnessed the reemergence of Ebola virus in Zaire causing a major outbreak among humans [30]. As of 24 August

**Table 1.** Medically important vector-borne virus infections and their occurrence

| Agent | Vector (s) | Occurrence |
|---|---|---|
| Dengue virus | *Aedes sp.* | Southeast Asia, Australia, Oceania, Africa, Carribbean, Central and South America |
| Chikungunya virus | *Aedes sp.* | Africa, Southeast Asia |
| Japanese encephalitis virus | *Culex sp., Aedes sp.* | Southeast Asia, Australia |
| Yellow fever virus | *Aedes sp.* | Africa, Central and South America |
| Rift Valley fever virus | *Culex sp., Aedes sp., Anopheles sp.* | Africa |
| West Nile virus | *Culex sp.* | Africa, Asia |
| Crimean-Congo haemorrhagic virus | *Hyalomma sp.* | Africa, Asia, Middle East, |
| Dugbe virus | *Hyalomma sp.* | Africa |
| Sindbis virus | *Culex sp.* | Africa, Asia, Australia, Eastern Europe |
| Sandfly fever virus | *Phlebotomus sp.* | Mediterranean, Central Asia |
| Ockelbo virus | *Culex sp.* | Northern Europe, Western Asia |
| Kyasanur Forest disease virus | *Haemaphysalis sp.* | India |
| Omsk haemorrhagic fever virus | *Dermacentor sp.* | Eastern Russia |
| Ross River virus | *Aedes sp.* | Australia, Oceania |
| Murray Valley Virus | *Culex sp.* | Australia |
| Mayaro virus | *Haemagogus sp.* | Central and South America |
| Oropouche virus | *Culicoides sp.* | South America |
| Venezuelan equine encephalitis virus | *Psorophora sp., Aedes sp., Mansonia sp.* | Central and South America |
| La crosse virus | *Aedes sp.* | North America |
| Western equine encephalitis virus | *Culex sp.* | North America |
| Eastern equine encephalitis virus | *Culex sp.* | North, Central and South America |
| St. Louis encephalitis virus | *Culex sp.* | North America |
| Colorado tick fever virus | *Dermacentor sp.* | North America |

**Table 2.** Medically important human rodent-borne virus infections
and their occurrence

| Agent | Rodent reservoir | Occurrence |
|---|---|---|
| Hantaan virus | *Apodemus sp.* | Asia |
| Puumala virus | *Clethrionomys sp.* | Europe, Asia |
| Balkan hantaviruses | *Apodemus sp.* | Southeast Europe |
| Seoul virus | *Rattus sp.* | Europe, Asia |
| Sin Nombre virus | *Peromyscus sp., Sigmodon sp.* | North America |
| Brasilian hantavirus | ? | Brazil |
| Guanarito virus | *Sigmodon sp.* | Venezuela |
| Junin virus | *Calomys sp.* | Argentina |
| Machupo virus | *Calomys sp.* | Bolivia |
| Sabia virus | ? | Brazil |
| Lassa virus | *Mastomys natalensis* | West Africa |

1995, a total of 315 human cases had occurred of which 244 (77%) had died [48].

## Prospects for the future

There are many vector- and rodent-borne infections endemic in countries frequently visited by travelers. An overview of the occurrence of vector- and rodent-borne virus is summarized in Tables 1 and 2. With the growing number of travelers to remote geographic regions on earth, there is a potential risk of acquiring vector- or rodent-borne infections. However, many clinicians in countries where these infections do not occur frequently, are unaware of this possibility. Laboratories should be prepared to offer diagnostic support for these infections. Some of these viruses, e.g. Crimean-Congo haemorrhagic fever virus, have a potential for nosocomial spread [13, 38], and public health officials and hospital hygiene consultants need to define guidelines to avoid secondary or nosocomial transmission of such viruses. Precautions are necessary to avoid the import of infected vectors, rodents or livestock by air or sea.

## References

1. Arthur RR, El-Sharkawy MS, Cope SE, Botros BA, Oun S, Morrill JC, Shope RE, Hibbs RG, Darwish MA, Imam IZE (1993) Recurrence of Rift Valley fever in Egypt. Lancet 342: 1149–1150
2. Calisher CH, Weinberg AN, Muth DJ, Lazuick JS (1987) Toscana virus infection in United States citizen returning from Italy. Lancet I: 165–166
3. Centers for Disease Control (1991) Imported dengue – United States, 1990. Morb Mort Wkly Rep 40: 519–520
4. Centers for Disease Control (1993) Imported dengue – United States, 1991. Dengue Surveill Summ 66: 1–4

5. Centers for Disease Control (1994) Dengue fever among U.S. military personnel – Haiti, September–November, 1994. Morb Mort Wkly Rep 43: 845–848

6. Centers for Disease Control and Prevention (1995) Imported dengue – United States, 1993–1994. Morb Mort Wkly Rep 44: 353–356

7. Cooper CB, Gransden WR, Webster M, King M, O'Mahony M, Young S, Banatvala JE (1982) A case of Lassa fever: experience at St. Thomas's Hospital. Br Med J 285: 1003–1005

8. Cunningham R, Mutton K (1991) Dengue haemorrhagic fever. Br Med J 302: 1083–1084

9. Deutman AF, Klomp HJ (1981) Rift Valley fever retinitis. Am J Ophthalmol 92: 38–42

10. Ehrnst A, Peters CJ, Niklasson B, Svedmayr A, Holmgren B (1985) Neurovirulent Toscana virus (a sandfly fever virus) in a Swedish man after visit to Portugal. Lancet I: 1212–1213

11. Eitrem R, Vene S, Niklasson B (1990) Incidence of sand fly fever among Swedish United Nations soldiers on Cyprus during 1985. Am J Trop Med Hyg 43: 207–211

12. Eitrem R, Niklasson B, Weiland O (1991) Sandfly fever among Swedish tourists. Scand J Infect Dis 23: 451–457

13. Fisher-Hoch SP, Khan JA, Rehman S, Mirza S, Khurshid M, McCormick JB (1995) Crimean Congo-haemorrhagic fever treated with oral ribavirin. Lancet 346: 472–475

14. Gaidamovich SV, Khutoretskaya NV, Asyamov YV, Tsyupa VI, Melnikova EE (1990) Sandfly fever in central Asia and Afganistan. In: Calisher CH (ed) Hemorrhagic fever with zonal syndrome, tick- and mosquito borne viruses. Springer, Wien New York, pp 287–293 (Arch Virol [Suppl] 1)

15. Gubler DJ, Trent DW (1993) Emergence of epidemic dengue/dengue hemorrhagic fever as a public health problem in the Americas. Infect Agents Dis 2: 383–393

16. Harnett GB, Bucens MR (1990) Isolation of chikungunya virus in Australia. Med J Aust 152: 328–329

17. Hasler C, Schnorf H, Enderlin N, Gyr K (1993) Importiertes Dengue-Fieber nach einem Tropenaufenthalt. Schweiz Med Wschr 123: 120–124

18. Hayes CG, O'Rourke TF, Fogelman V, Leavengood DD, Crow G, Albersmeyer MM (1989) Dengue fever in American military personnel in the Philippines: clinical observations on the hospitalized patients during a 1984 epidemic. Southeast Asian J Trop Med Pub Health 20: 1–8

19. Hirabayashi Y, Oka S, Goto H, Shimada K, Kurata T, Fisher-Hoch SP, McCormick JB (1988) An imported case of Lassa fever with late appearance of polyserositis. J Infect Dis 158: 872–875

20. Holmes GP, McCormick JB, Trock SC, Chase RA, Lewis SM, Mason CA, Hall PA, Brammer LS, Perez-Oronoz GI, McDonnell MK, Paulissen JP, Schonberger LB, Fisher-Hoch S (1990) Lassa fever in the United States – investigation of a case and new guidelines for management. N Engl J Med 323: 1120–1123

21. Jacobs MG, Brook MG, Weir WRC, Bannister BA (1991) Dengue haemorrhagic fever, a risk of returning home. Br Med J 302: 828–829

22. Jahrling PB, Geisbert TW, Dalgard DW, Johnson ED, Ksiazek TG, Hall WC, Peters CJ (1990) Preliminary report: isolation of Ebola virus from monkeys imported to USA. Lancet 335: 502–505

23. Kanesa-Thasan N, Iacono-Connors L, Magill A, Smoak B, Vaughn D, Dubois D, Burrous J, Hoke C (1994) Dengue serotype 2 and 3 in US forces in Somalia. Lancet 343: 678 (1994)

24. Kenyon RH, Niklasson B, Jahrling PB, Geisbert T, Svensson L, Frydén A, Bengtsson M, Foberg U, Peters CJ (1994) Virologic investigation of a case of suspected haemorrhagic fever. Res Virol 145: 397–406

25. Krippner R, Hanisch G, Kretschmar H (1990) Denguefieber mit hämorrhagischen Manifestationen nach Thailandaufenthalt. Dtsch Med Wochenschr 115: 858–862
26. Le Guenno B, Formentry P, Wyers M, Gounon P, Walker F, Boesch C (1995) Isolation and partial characterisation of a new strain of Ebola. Lancet 345: 1271–1274
27. Mahdy MS, Chiang W, McLaughlin B, Derksen K, Truxton BH, Neg K (1989) Lassa fever: the first confirmed case imported into Canada. Can Dis Wkly Rep 15: 193–198
28. Macdonald WBG, Tink AR, Ouvrier RA, Menser MA, de Silva LM, Naim H, Hawkes RA (1989) Japanese encephalitis after a two-week holiday in Bali. Med J Aust 150: 334–339
29. Morens DM, Sather GE, Gubler DJ, Rammohan M, Woodall JP (1987) Dengue shock syndrome in an American traveler with primary dengue 3 infection. Am J Trop med hyg 36: 424–426
30. Muyembe T, Kipasa M (1995) Ebola haemorrhagic fever in Kitwit, Zaire. Lancet 345: 1448
31. Niklasson B, Meegan JM, Bengtsson E (1979) Antibodies to Rift Valley fever virus in Swedish U.N. soldiers in Egypt and the Sinai. Scand J Infect Dis 11: 313–314
32. Patey O, Ollivaud L, Breuil J, Lafaix C (1993) Unusual neurologic manifestations occurring during dengue fever infection. Am J Trop Med Hyg 48: 793–802
33. Schwarz TF, Gilch S, Jäger G (1993) Travel-related Toscana virus infection. Lancet 342: 803–804
34. Schwarz TF, Jäger G, Gilch S, Pauli C (1995) Serosurvey and laboratory diagnosis of imported sandfly fever virus, serotype Toscana, infection in Germany. Epidemiol Infect 114: 501–510
35. Schwarz TF, Gilch S, Jäger G (1995) Aseptic meningitis caused by sandfly fever virus, serotype Toscana. Clin Infect Dis 21: 669–671
36. Schwarz TF, Jäger G (1995) Imported dengue virus infections in German tourists. Zbl Bakteriol 282: 533–536
37. Siegert R, Shu HL, Slenczka W, Peters D, Müller G (1967) Zur Ätiologie einer unbekannten von Affen ausgegangenen Infektionskrankheit. Dtsch Med Wochenschr 92: 2341–2343
38. Suleiman MNEH, Muscat-Baron JM, Harries JR, Satti AGO, Platt GS, Bowen ETW, Simpson DIH (1980) Congo/Crimean haemorrhagic fever in Dubai – an outbreak at the Rashid hospital. Lancet II: 939–941
39. Swanepoel R (1987) Viral haemorrhagic fevers in South Africa: history and national strategy. S Afr J Sci 83: 80–88
40. Van Tongeren HAE (1981) Imported virus diseases in the Netherlands out of tropical areas 1977–1980 (30 months). Tropenmed Parasit 32: 205
41. Vaughn DW, Hoke CH (1992) The epidemiology of Japanese encephalitis: prospects for prevention. Epidemiol Rev 14: 197–221
42. Vögtlin J, Gyr K (1985) Dengue-Fieber als Importkrankheit in der Schweiz. Schweiz Med Wochenschr 115: 1273–1277
43. Wittesjö B, Eitrem R, Niklasson B (1993) Dengue fever among Swedish tourists. Scand J Infect Dis 25: 699–704
44. Wittesjö B, Eitrem R, Niklasson B, Vene S, Mangiafico JA (1995) Japanese encephalitis after a 10-day holiday in Bali. Lancet 345: 856
45. Woodruff AW, Bowen ET, Platt GS (1978) Viral infections in travellers from tropical Africa. Br Med J I: 956–958
46. World Health Organization (1992) Viral hemorrhagic fever in imported monkeys. Wkly Epidemiol Rep 67: 142

47. World Health Organization (1995) Japanese encephalitis. Wkly Epidemiol Rep 70: 166–167
48. World Health Organization (1995) Ebola haemorrhagic fever. Wkly Epidemiol Rep 70: 241–242
49. Zweighaft RM, Fraser DW, Hattwick MAW, Winkler WG, Jordan WC, Alter M, Wolfe M, Wulff H, Johnson KM (1977) Lassa fever: response to an imported case. N Engl J Med 297: 803–807

Author's address: Dr. T. F. Schwarz, Stiftung Juliusspital, Central Laboratory, Juliuspromenade 19, D-97070 Würzburg, Federal Republic of Germany.

Arch Virol (1996) [Suppl] 11: 13–20

# WHO program on emerging virus diseases

**J. W. LeDuc**

Division of Communicable Diseases, World Health Organization,
Geneva, Switzerland

**Summary.** Infectious diseases, and especially viral diseases, are important, evolving, complex public health problems. Their ultimate prevention and control will increasingly require sophisticated interaction between epidemiologic resources, molecular expertise, and application of modern statistical tools. The integration of epidemiologic and laboratory sciences is central to the success of a coordinated approach to new, emerging and re-emerging infectious diseases, and the WHO is attempting to facilitate and strengthen these resources internationally by focusing on improved surveillance, infrastructure building, applied research, and improved prevention and control strategies. Reference virus laboratories, especially those that deal with exotic virus diseases, will play a key role in implementation of the program, since these laboratories are likely to be called upon to assist in the identification of new, emerging, or re-emerging diseases. Thus, ensuring that these laboratories are well prepared to perform their critical tasks is essential to the ultimate success of the program.

## Introduction

In this book detailed descriptions will be presented of various important viruses and the diseases they cause. Some of these are well known pathogens, while others, like recently recognized South American arenaviruses and Sin Nombre virus, the cause of hantavirus pulmonary syndrome, are new and justifiably called emerging virus diseases. This paper will not refer to any single virus or disease, but rather deal in more general terms with the growing public health problem of emerging infectious diseases, and describe some of the initiatives now under way at the World Health Organization (WHO) in our attempts to address this important topic.

In April 1994, the WHO held an organizational meeting to address the issue of emerging infectious diseases. Attention was brought to this subject by two significant publications. The United States Institute of Medicine (IOM) report, *Emerging Infections* [1], was published in 1992, and crystallized the thinking of many scientists and public health officials who had been concerned about the growing incidence of new diseases, most notably HIV and AIDS, but also the dramatic increase of other diseases such as dengue, and the threatening problem of antimicrobial resistance. This book targeted public health issues in the United States, but in fact, the vast majority of problems discussed were international in nature, and applied equally well to most other countries of the world.

In response to the challenges raised in the Institute of Medicine report, the US Centers for Disease Control and Prevention (CDC) developed a detailed plan entitled *Addressing Emerging Infectious Disease Threats, a Prevention Strategy for the United States* [2]. This plan listed specific goals and tasks, which are now being systematically addressed. As with the Institute of Medicine report, the CDC plan, while focused on US public health issues, raised many points that are equally important internationally, and in fact a significant component of the plan involves international activities related to emerging infectious diseases.

At about the time the CDC plan was released, WHO hosted its first meeting on emerging infectious diseases. The objective was to lay the ground work necessary for an international program on emerging infectious diseases, and both the IOM and CDC reports weighted heavily in the design and direction of the final WHO plan [3].

## WHO plan objectives

Four specific goals were adopted during the WHO meeting [3]. The first was to strengthen global surveillance of infectious diseases, and there the call was made to define existing networks of WHO Collaborating Centers that could be used to recognize, report and respond to specific infectious disease problems. For example, networks of collaborating centers are already in place to investigate influenza, polio, HIV/AIDS and arboviruses and haemorrhagic fevers. Antimicrobial resistance was also recognized as a critical target for surveillance, and it was recommended that special efforts be made to monitor antimicrobial resistance trends globally. With the dramatic changes underway worldwide in food production and distribution, special attention was also drawn to zoonotic diseases and foodborne pathogens.

The second goal was to strengthen national infrastructures necessary to recognize and respond to emerging infectious diseases. To do this, we hope to build laboratory capabilities, increase training opportunities, and improve communications between national, regional and international resources.

The need was recognized to use the advances of the biotechnological revolution to address practical applications in infectious disease diagnosis, especially in the production of accurate, inexpensive tests suitable for use in developing countries. We also need to better understand the epidemiology of various diseases so that improved strategies of disease prevention can be made. Further, we need to foster regional collaborations, especially directed towards regional self-sufficiency in reagents production and quality control. Finally, there is a need to evaluate and set standards for basic public health practices. Thus, the third goal was to encourage an applied research program.

Last, the need was recognized to strengthen prevention and control strategies by developing specific guidelines for various known disease conditions such as zoonoses, foodborne pathogens, and others, and to actively intervene to prolong the useful lifespan of antimicrobial agents. Also, we need to develop better ways of exchanging health information.

## Implementation of the WHO plan

To accomplish these goals, we set out to review those critical elements necessary for effective surveillance and response to emerging infectious diseases. Clearly the focus of our work will be to deal with human pathogens, so we must have access to acutely ill patients. Further, to understand why these patients are sick, we must have laboratory-based diagnostic capabilities, and these laboratories must be able to accurately confirm common pathogens present in their region, so that they can recognize what is new or unusual. Further, we will require some form or reference facility for assistance and confirmation of difficult diagnoses. We also need the ability to communicate rapidly and reliably, and we will require a centralized facility to receive information, coordinate efforts and assist in response activities when problems are identified. We should also have the capacity to train and motivate collaborators, and a focal point to lead fund-raising activities.

Given these prerequisites, it became clear that the most efficient means of approaching this problem was to make greater use of the existing network of WHO Collaborating Centers, especially those that deal with exotic virus diseases, since they are the ones most often called upon to assist in the identification of unknown agents. Most of these centers have access to acutely ill patients, they have laboratory-based diagnostics, there is a referral laboratory system in place, and they could be actively coordinated through WHO Headquarters.

We then set about to develop a plan of action for this program. Our short-term goals were to assess the capabilities of the existing networks of Collaborating Centers, identify those eager to participate in this initiative, and begin frequent contact with and between them.

We next hoped to finalize the composition of the network, identify critical diagnostic reagent needs, improve communications, and enhance WHO's ability to respond to emergencies.

Our long term goals are then to develop a dynamic, interactive global network of Collaborating Centers able to readily differentiate routinely encountered diseases from those that might be considered new, emerging, or re-emerging, and to investigate and report the appearance of such diseases as they are recognized. We also hope that this network will allow a rapid exchange of technology, equipment advances, and reagents. This network should also be able to assist with training activities of both students and professionals, and to have ready access to a communications network for prompt information exchange.

## Survey of WHO virus collaborating centers capabilities

With these goals and plan of action in mind, in 1993 we surveyed the existing network of Collaborating Centers and selected other laboratories with expertise in arboviruses and haemorrhagic fevers. The 34 laboratories participating in this survey are listed in Table 1. We first asked if the laboratories were associated with their national Ministry of Health, if they had immediate access to clinical material through a direct association with a hospital, if they maintained a serum bank, and the kinds of techniques available to them for virus isolation. The results of these questions are summarized in Table 2 and indicate that the majority of participating laboratories responded positively to these questions. We then ask about the serological techniques in use in their laboratories, and most laboratories were found to be well equipped for serological analyses. Next we asked what kinds of molecular techniques were routinely used in their facility, and again, the laboratories were in general using modern molecular techniques routinely (Table 2).

When each laboratory was asked if they had the reagents necessary to diagnose a human infection caused by an alphavirus, less than half had the reagents necessary to do so (Fig. 1). Similarly, when questioned about common flavivirus infections such as yellow fever and dengue, about two-thirds had the necessary reagents to make an accurate diagnosis. Reagents were even further limiting in terms of the bunyaviruses, with less than half the laboratories equipped to diagnose infectious due to these viruses. And the situation was worst with the causes of viral haemorrhagic fever, where only about a quarter of the labs had the necessary reagents to confirm an infection due to these viruses (Fig. 1).

The conclusion of the survey was then that the laboratory network was generally well prepared to participate in global surveillance of new, emerging and re-emerging infectious diseases in terms of technical skills and access to acutely ill patients, but most laboratories lacked a complete set of reagents necessary to recognize even common viral infections that they may encounter, especially for those diseases not frequently seen in their particular area, as might be the case when examining clinical specimens from returning travellers.

In response to this identified weakness in the proposed network, we have actively begun production and distribution of critical diagnostic reagents for provision to these laboratories. We have started with antigens to dengue, yellow fever and Rift Valley fever viruses. We are also attempting to collect human IgM positive sera from these and other important viral diseases for use as positive control reagents. This has been especially important in the case of yellow fever and dengue. Finally, we are working closely with the WHO Expanded Program on Immunizations and their polio eradication laboratory network and the training initiative to include laboratory diagnosis for yellow fever as part of the training sessions. Most activities have been centered in Africa, and we have completed two workshops already, with a third scheduled soon. In addition, we held a special workshop in Kenya in June 1995, specifically dedicated to the laboratory diagnosis of yellow fever and other arbovirus and haemorrhagic fevers.

New, emerging and re-emerging infectious diseases are not limited only to those caused by viruses. Consequently, brief mention is appropriate regarding

**Table 1.** World Health Organization Collaborating Centers and other reference laboratories participating in a 1993 survey of technical capacity to address new, emerging, and re-emerging virus diseases

| Laboratory | Location |
| --- | --- |
| The Americas | |
| Laboratory Center for Disease Control | Ottawa, Canada |
| Special Pathogens Branch, Centers for Disease Control and Prevention | Atlanta, Georgia, USA |
| Division of Vector-Borne Infectious Diseases Centers for Disease Control and Prevention | Fort Collins, Colorado, USA |
| Yale Arbovirus Research Unit | New Haven, Connecticut, USA |
| Instituto "Pedro Kouri" | Havana, Cuba |
| Dengue Branch Centers for Disease Control and Prevention | San Juan, Puerto Rico |
| Instituto Evandro Chagas | Belem, Brazil |
| Instituto Adolfo Lutz | São Paulo, Brazil |
| Instituto Nacional de Enfermedades Virales Humanos | Pergamino, Argentina |
| South-East Asia | |
| National Institute of Virology | Pune, India |
| Centre for Vaccine Development | Nakhon Pathom, Thailand |
| Western Pacific | |
| University of Western Australia | Perth, Australia |
| Queensland Institute of Technology | Brisbane, Australia |
| Institute of Tropical Medicine | Nagasaki, Japan |
| National Institute of Health | Tokyo, Japan |
| University of Malaya | Kuala Lumpur, Malaysia |
| Korea University | Seoul, Korea |
| Europe | |
| Institute Pasteur | Paris, France |
| Institute of Molecular Genetics | Heidelberg, Germany |
| Aristotelian University | Thessaloniki, Greece |
| Istituto Superiore de Sanita | Rome, Italy |
| Erasmus University | Rotterdam, Netherlands |
| Institute of Virology | Bratislava, Slovak Republic |
| Medical Faculty of Ljubljana | Ljubljana, Slovenia |
| Centre for Applied Microbiology and Research | Porton Down, Salisbury, UK |
| Swedish Institute for Infectious Disease Control | Stockholm, Sweden |
| Institute of Poliomyelitis and Viral Encephalitides | Moscow, Russian Federation |
| Ivanovsky Institute of Virology | Moscow, Russian Federation |
| Africa | |
| Pasteur Institute | Bangui, Central African Republic |
| Pasteur Institute | Dakar, Senegal |
| Kenya Medical Research Institute | Nairobi, Kenya |
| University of Ibadan | Ibadan, Nigeria |
| Uganda Virus Research Institute | Entebbe, Uganda |
| National Institute for Virology | Sandringham, South Africa |

**Table 2.** Summary of WHO Collaborating Center's associations
and technical resources

| | |
|---|---|
| Associations | |
| Directly associated with national Ministry of Health | 74%[a] |
| Directly associated with a hospital or clinical care facility | 59% |
| Technical resources | |
| Maintain cell cultures | 100% |
| Access to suckling mice | 85% |
| Access to mosquitoes for virus isolation | 32% |
| Serological techniques used | |
| Enzyme immunoassay | 100% |
| Immunofluorescent assay | 100% |
| Hemagglutination inhibition | 91% |
| Complement fixation test | 81% |
| Radioimmunoassay | 44% |
| Plaque reduction neutralization test | 94% |
| Agar gel immunodiffusion | 56% |
| Mouse neutralization test | 74% |
| Molecular techniques used | |
| Western blot | 91% |
| Polyacrylamide gel electrophoresis | 85% |
| Polymerase chain reaction | 91% |
| Ultracentrifugation capability | 94% |
| Protein purification capability | 79% |
| Radio-labelling capability | 59% |
| Sequencing capability | 71% |
| Cloning capability | 65% |
| Expression capability | 47% |

[a]Percentage of laboratories indicating a positive response. Not all laboratories responded to all inquiries

some other activities now in progress on emerging diseases at WHO. First, the Director-General has proposed a resolution to the Executive Board of the World Health Assembly recognizing the importance of emerging infectious diseases, and stating the four goals outlined above as the focus of the WHO effort in this field. The resolution urges Member States to strengthen national and local surveillance programs, to improve routine diagnostic capabilities, to enhance communications, to foster more applied research and establish prevention strategies, and specifically calls for greater concern about antimicrobial sensitivity and to encourage more rational use of existing antimicrobial agents. The resolution also calls for efforts to strengthen WHO's capacity to address issues of emerging infectious diseases, and encourages greater coordination of the various

**WHO Collaborating centres for arboviruses and haemorrhagic fevers**

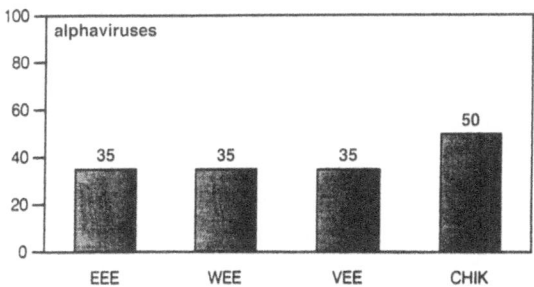

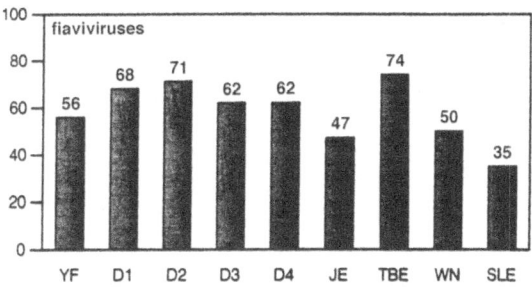

**Fig. 1.** Percentage of laboratories indicating availability of diagnostic reagents for the viruses are indicated. (Alphaviruses: *EEE* eastern equine encephalomyelitis; *WEE* Western equine encephalomyelitis; *VEE* Venezuelan equine encephalomyelitis; *CHIK* Chikungunya; Flaviviruses: *YF* Yellow Fever; *D1* dengue 1; *D2* dengue 2; *D3* dengue 3; *D4* dengue 4; *JE* Japanese encephalitis; *TBE* tick-borne encephalitis (any type); *WN* West Nile; *SLE* St. Louis Encephalitis; Bunyaviruses; *CCHF* Crimean Congo haemorrhagic fever; *RVF* Rift Valley fever; *SF* Sandfly fever (any type); *ORO* Oropouche; *CAL* California group (any type); *HTN* Hantaan. Arenaviruses and Filoviruses: *LAS* Lassa; *LCM* lymphocytic choriomeningitis; *NWarenas* New World arenaviruses (any type); *MAR* Marburg; *EBO* Ebola; *RST* Reston)

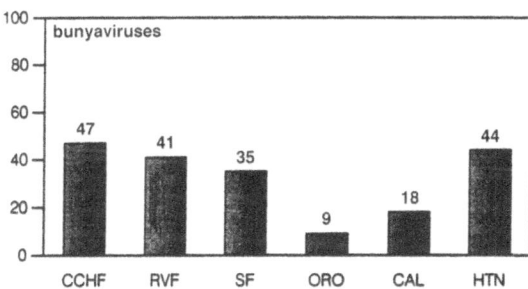

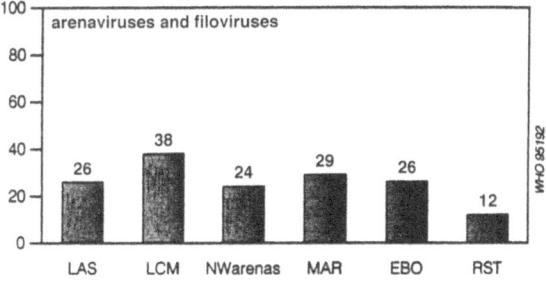

technical programs within WHO to develop this initiative. The resolution was endorsed by the Executive Board of the World Health Assembly, and approved by the full Assembly during their annual meeting in May, 1995.

To implement this program, we have organized an internal coordinating group, which meets periodically to discuss areas of mutual interest in emerging infectious diseases. Each collaborating group prepares a summary of those activities they consider to be appropriate for consideration under new and emerging diseases, and these are then arranged together for submission to donor agencies in hopes of obtaining extrabudgetary support.

## References

1. Lederberg J, Shope RE, Oaks SC (eds) Emerging infections, microbial threats to health in the United States. Institute of Medicine, National Academy Press, Washington
2. Centers for Disease Control and Prevention (1994) Addressing emerging infectious disease threats: a prevention strategy for the United States. Atlanta, Georgia, U.S. Department of Health and Human Services, Public Health Service
3. World Health Organization (1994) Emerging infectious diseases: memorandum from a WHO meeting. Bull World Health Organ 72: 845–850

Author's address: Dr. J. W. LeDuc, Division of Communicable Diseases, World Health Organization, Geneva, Switzerland.

Arch Virol (1996) [Suppl] 11: 21–32

# Arboviruses as imported disease agents: the need for increased awareness

**D. J. Gubler**

Division of Vector-Borne Infectious Diseases, National Center for Infectious Diseases, Centers for Disease Control and Prevention, Public Health Service, U.S. Department of Health and Human Services, Fort Collins, Colorado, U.S.A.

**Summary.** The arboviruses are an important group of etiologic agents that are transported between geographic regions in infected animals and humans. This group of viruses is briefly reviewed as agents of imported disease and dengue viruses are discussed as an example to illustrate the trend of increasing incidence of imported arboviral diseases.

## Global distribution of arboviruses

The arboviruses are a heterogeneous group that require a hematophagous arthropod for transmission from one vertebrate host to another. All true arboviruses produce a significant viremia in their natural vertebrate host. The arthropod is exposed to and becomes infected with the virus when it takes a blood meal from the viremic host. The major arthropod vectors of arboviruses include mosquitoes, sandflies, midges, and hard and soft ticks; the major vertebrate hosts are rodents and birds. A characteristic of these viruses is their natural association with specific vertebrate hosts and arthropod vectors. Over 535 arboviruses and zoonotic viruses have been described in 13 taxonomic families; many of the latter are rodent-borne viruses and do not require an arthropod vector [15]. The true arboviruses belong to seven taxonomic families, but most of those causing human disease belong to only three of these (*Togaviridae, Flaviviridae,* and *Bunyaviridae,* Table 1).

The majority of arboviruses are maintained in natural cycles involving a nonhuman primary vertebrate host and a primary arthropod vector (Fig. 1). Most such cycles remain unknown until humans encroach on the natural enzootic focus, or the virus escapes the primary cycle via a secondary vector or vertebrate host as the result of some ecologic change. Humans and domestic animals generally become involved only after the virus is brought into the peridomestic environment by a bridge vector; they frequently develop clinical illness and are "dead-end" hosts because they do not produce viremia and, therefore, do not contribute to the transmission cycle.

**Table 1.** The more important arboviruses causing human disease

| Family virus | Vector | Vertebrate host | Ecology[b] | Disease in humans | Geographic distribution | Epidemics |
|---|---|---|---|---|---|---|
| *Togaviridae* | | | | | | |
| Chikungunya[a] | mosquitoes | man, primates | U, S, R | SFI[c] | Africa, Asia | yes |
| Ross River[a] | mosquitoes | man, marsupials | R, S, U | SFI | Australia, So. Pacific | yes |
| Mayaro | mosquitoes | birds | R, S, U | SFI | South America | yes |
| O'nyong-nyong | mosquitoes | ? | R | SFI | Africa | yes |
| Sinbis | mosquitoes | birds | R | SFI | Asia, Africa, Australia, Europe | yes |
| Eastern Equine encephalitis | mosquitoes | birds | R | SFI, ME | Americas | yes |
| Western Equine encephalitis | mosquitoes | birds, rabbits | R | SFI, ME | Americas | yes |
| Venezuelan Equine encephalitis[a] | mosquitoes | rodents | R, S | SFI, ME | Americas | yes |
| Barmah Forest[a] | mosquitoes | rodents | R | SFI | Americas | yes |
| *Flaviviridae* | | | | | | |
| Dengue 1–4[a] | mosquitoes | man, primates | U, S, R | SFI, HF | worldwide in tropics | yes |
| Yellow Fever[a] | mosquitoes | man, primates | R, S, U | SFI, HF | Africa, South America | yes |
| Kyasanur Forest disease | mosquitoes | primates, rodents | R | SFI, HF, ME | India | no |
| Omsk hemorrhagic fever | ticks | rodents | R | SFI, HF | Asia | no |
| Japanese encephalitis | mosquitoes | birds | R, S | SFI, ME | Asia | yes |
| Murray Valley encephalitis | mosquitoes | birds | R | SFI, ME | Australia | yes |
| Rocio | mosquitoes | birds | R | SFI, ME | South America | yes |
| St. Louis encephalitis | mosquitoes | birds | R, S, U | SFI, ME | Americas | yes |
| West Nile | mosquitoes | birds | R, S, U | SFI, ME | Asia, Africa | yes |
| Tick-borne encephalitis | ticks | rodents | R | SFI, ME | Europe, Asia | no |
| *Bunyaviridae* | | | | | | |
| Sandfly fever[a] | sandflies | | R | SFI | Europe, Africa, Asia | yes |
| Rift Valley fever[a] | mosquitoes | | R | SFI, HF, ME | Africa | yes |
| La Crosse encephalitis | mosquitoes | rodents | R, S | SFI, ME | North America | no |
| California encephalitis | mosquitoes | rodents | R | SFI, ME | North America, Europe, Asia | yes |
| Congo-Crimean hemorrhagic fever[a] | ticks | rodents | R | SFI, HF | Europe, Asia, Africa | yes |
| Oropouche[a] | midges | | R, S, U | SFI | Central & South America | yes |

[a] Arboviruses that produce significant human viremia
[b] *U* Urban; *S* suburban; *R* rural; underline designates the most important ecology
[c] *SFI* System febrile illness; *ME* meningoencephalitis; *HF* hemorrhagic fever

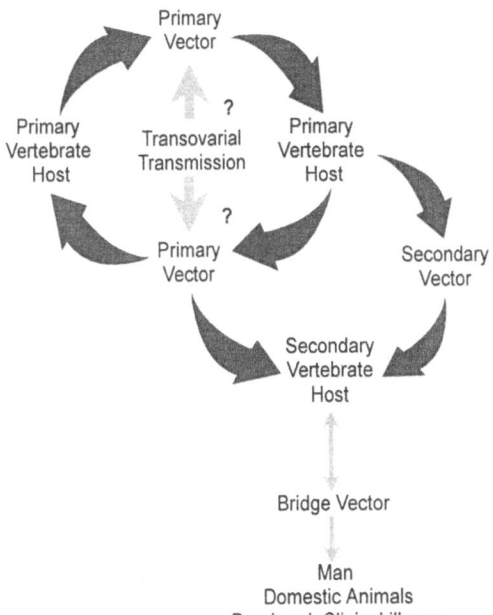

**Fig. 1.** Generalized arbovirus maintenance cycle

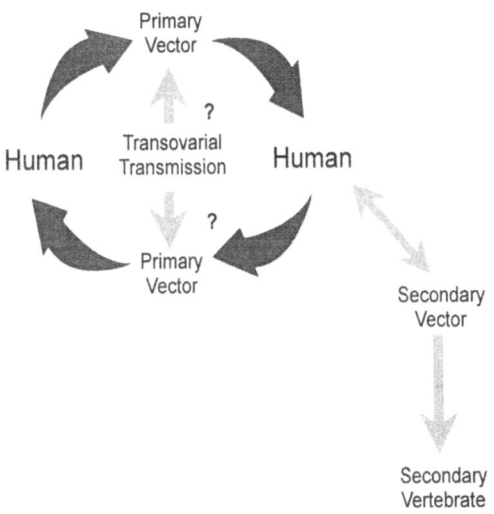

**Fig. 2.** Epidemic/maintenance arbovirus cycle involving humans as the primary vertebrate host

A few arboviruses cause significant viremia in humans and may be transmitted in a human-arthropod-human cycle (Fig. 2). This is generally an epidemic transmission cycle, but with some viruses such as dengue, it may also be an endemic maintenance cycle [8]. Occasionally, a secondary vector may transmit the virus to other vertebrates, which are usually dead-end hosts. In both types of transmission cycles, maintenance of the viruses in nature may be facilitated by transovarial transmission, in which the virus is transmitted from the female through the eggs to the offspring. The significance of transovarial transmission or whether it occurs is not known for most arboviruses.

Arboviruses have a global distribution, but the majority are found in tropical developing countries [15]. The greatest number of viruses have been described from Africa and South America, where the flora and fauna are diverse and extensive studies have been conducted. The smaller number of viruses described from Asia, which has ecologic diversity comparable to Africa and tropical America, may simply reflect the lack of studies conducted in that region. The geographic distribution of each arbovirus is limited by the ecologic parameters governing its transmission cycle. Important limiting factors include temperature, rainfall patterns, distribution of the arthropod vector and vertebrate reservoir host.

The majority of arbovirus infections in humans result in a nonspecific viral syndrome [24]. Onset is usually sudden with fever, headache, myalgias, malaise and occasionally prostration. Infection with some arboviruses may lead to more severe hemorrhagic disease or encephalitis, often with a fatal outcome or permanent neurologic sequelase. Table 1 lists the most important arboviruses causing human disease along with their arthropod vector, the most common vertebrate hosts, the type of illness in humans, their geographic distribution and whether or not virus transmission is associated with epidemics. It will be noted that mosquitoes are the most common arthropod vector of human disease; birds are the most important reservoir hosts, although rodents and lower primates are also important. Dengue viruses have the widest geographic distribution. They are found worldwide in the tropics (Fig. 3), and are by far the most common as

▨ Areas infested with *Aedes aegypti* and at risk for epidemic dengue
▦ Countries with recent dengue activity

**Fig. 3.** The world distribution of dengue viruses and their primary mosquito vector, *Aedes aegypti*, in 1995

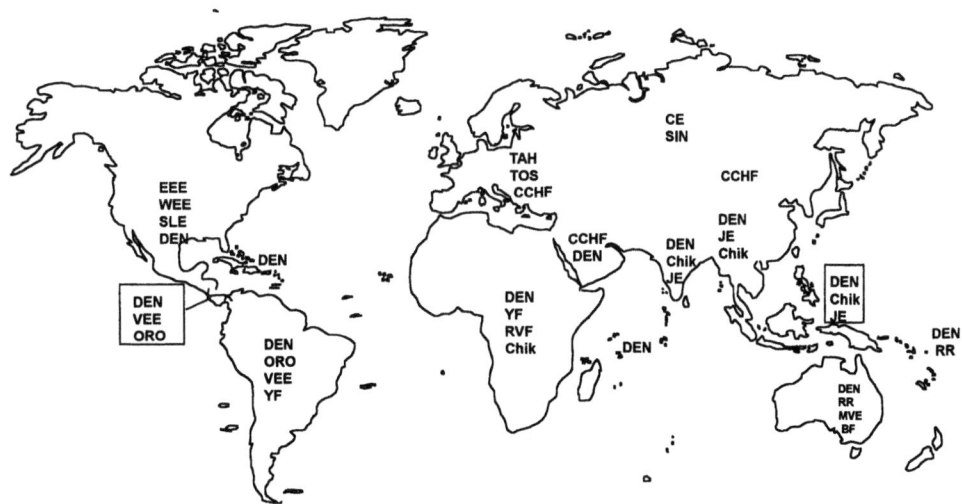

**Fig. 4.** Global distribution of arbovirus epidemics, 1980–1995. *BF* Barmah Forest, *CE* California Encephalitis, *Chik* Chikungunya, *CCHF* Congo-Crimean Hemorrhagic Fever, *DEN* Dengue, *EEE* Eastern Equine Encephalitis, *JE* Japanese Encephalitis, *MVE* Murray Valley Encephalitis, *ORO* Oropouche, *RVF* Rift Valley Fever, *RR* Ross River, *SLE* St. Louis Encephalitis, *TAH* Tahyna, *TOS* Toscana, *VEE* Venezuelan Equine Encephalitis, *WEE* Western Equine Encephalitis, *YF* Yellow Fever

a cause of human disease [8, 10]. Other viruses such as Sinbis have a widespread geographic distribution, but are less important as a known cause of human disease. Some viruses, such as Kyasanar Forest virus in Karnataka State in India, have a very limited focal distribution. Arboviruses are found in diverse ecologic settings, but most human pathogens occur in rural and urban/suburban areas. For most, exposure in the latter areas is a problem only during epidemic transmission when the virus has been brought into the peridomestic environment by a bridge vector. Dengue viruses are exceptions, as they are maintained in urban areas of the tropics [8].

There has been a significant emergence of epidemic arboviral disease worldwide in the past 15 years [10]. Each of the major geographic regions has experienced large epidemics of at least one arbovirus during this time, and those in the tropics and subtropics have experienced multiple epidemics of several arboviruses (Fig. 4). The factors responsible for emergence vary with each virus. For example, there has been a dramatic increase in both the incidence and a geographic expansion of the dengue viruses associated with urbanization of tropical developing countries. This will be discussed in more detail below. On the other hand, the emergence of diseases like Rift Valley fever in Africa, Japanese encephalitis in Asia and Oropouche in Central and South America are all associated with changing agricultural or irrigation practices [10]. For each of these viruses there are specific changes in the ecology that resulted in humans having more frequent contact with either the primary or secondary transmission cycles, thus increasing the risk of human exposure. The asterisked viruses (Table 1) cause significant viremia in humans; most are transmitted by mosquitoes.

Because most arboviral infections result in a nonspecific febrile illness, it is difficult to clinically differentiate them from a wide variety of other viral, bacterial and parasitic infections [24]. To diagnose these infections accurately, three things are essential: a detailed clinical summary, detailed epidemiologic data on the possible circumstances of infection (including a travel history that lists the dates and places the person has visited in the past month and information on the ecology of those areas), and a diagnostic laboratory test to make a definitive diagnosis. Both serologic and virologic methods are used for laboratory diagnosis.

Those arbovirus infections of humans (dengue, yellow fever, Chikungunya, Ross River, Venezuelan equine encephalitis, Rift Valley fever, Congo-Crimean hemorrhagic fever, Oropouche and other viruses) that produce a high viremia in humans, can be isolated from the blood of acutely ill patients. A number of mammalian and mosquito cell lines, baby mice or intrathoracic inoculation of mosquitoes can be used to isolate arboviruses from serum, cerebral spinal fluid and tissue samples [4, 12, 21, 25].

Definitive serologic diagnosis of arboviral infections requires paired acute and convalescent serum samples to demonstrate a seroconversion (four-fold or greater change in antibody titer) to specific viral antigens. A number of tests can be used to measure antibodies including the hemagglutination-inhibition, the complement-fixation, the neutralization, immunofluorescent assay and the enzyme-linked immunosorbent assay for IgG and the enzyme-linked immunosorbent and the immunofluorescent assays for IgM [4, 12, 25]. Many arboviruses are closely related to each other antigenically. As a result, there may be considerable cross-reactivity in serologic tests, making identification of the infecting virus difficult when only serologic methods are used. A detailed discussion of laboratory diagnostic tests is beyond the scope of this paper.

### Imported arbovirus diseases

The movement of people and commodities between countries and regions via jet airplane has increased dramatically in recent years. For example, the annual number of air passengers departing from the United States has doubled since 1983, from nearly 20 million to an estimated 40 million in 1994 [6] (Fig. 5). In most years, over 50% of these passengers traveled to a tropical destination, thus increasing potential exposure to vector-borne diseases. Rapid air travel provides the ideal mechanism for the transport of disease agents, their reservoir hosts and arthropod vectors to new geographic regions. This type of rapid movement is especially important for those mosquito-borne arboviruses that produce viremia in humans (Table 1). Moreover, because of the large numbers of visitors to tropical regions of the world, there is increased exposure to a variety of other arboviral diseases, such as Japanese encephalitis, that cause serious illness in

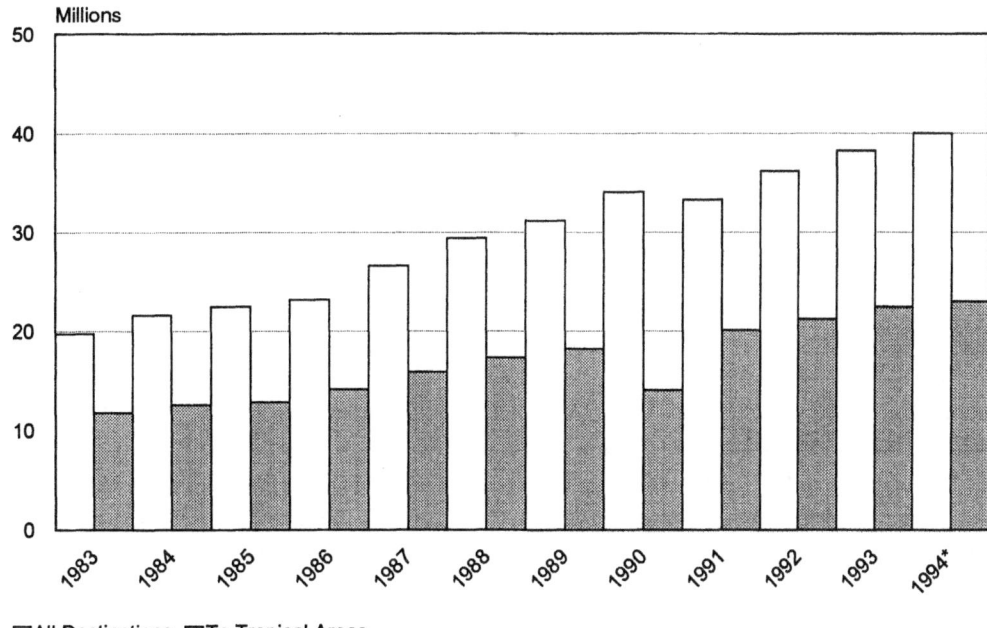

Millions

□All Destinations ▨To Tropical Areas

**Fig. 5.** The number of air passengers departing from the United States, 1983–1994. Source: U.S. Department of Transportation, IP92 data. Project based on data from three quarters

humans, but do not pose a threat to the populations in temperate regions [29].

Dengue is the most important arboviral disease of humans and will be discussed in more detail to underscore the importance of exotic viruses to travelers. Increased air travel has resulted in constant and repeated introductions of new dengue virus strains and serotypes into areas of the world where the principal mosquito vector, *Aedes aegypti*, occurs. The result has been increased frequency of epidemic dengue fever, expanding geographic distribution of both the vector and the viruses, the development of hyperendemicity (the co-circulation of multiple virus serotypes in a community) and the emergence of the severe and fatal form of disease, dengue hemorrhagic fever (DHF) [13]. The emergence of dengue/DHF as a major epidemic disease of the tropics has made it a global public health problem because even in temperate areas where endemic dengue transmission does not occur, the disease is being seen more often in travelers returning from tropical endemic areas.

In the United States, surveillance for imported dengue is passive and relies on the attending physician to make the diagnosis, take appropriate samples for testing in the laboratory and report the case to the state health department. Unfortunately, physicians in the United States have a low index of suspicion for dengue fever because the disease is not endemic. Therefore, they rarely include dengue in the differential diagnosis of viral syndrome, even in those patients with a travel history. As a result, many imported cases go unreported. Even so, over

2265 suspected cases of dengue infection have been reported to the Centers for Disease Control and Prevention (CDC) from 1977 through 1994. Of these, appropriate blood samples were taken on only a portion. Nevertheless, 498 (22%) were laboratory confirmed as dengue infections [5, 20]. Cases were caused by all four virus serotypes and were introduced from all major regions of the tropics. The number of imported cases to the United States has been decreasing in recent years, but this most likely reflects a trend of decreased emphasis on the passive dengue surveillance system in the continental United States. With waning resources for surveillance in recent years, emphasis has been placed on disease-endemic areas, where the disease is a major public health problem. The percentage of cases confirmed has remained fairly constant over the years, but increased dramatically in 1994. This was most likely the result of widespread epidemic transmission in the Caribbean in 1994.

Of interest is that on two occasions recently (1980 and 1986), autochthonous dengue transmission occurred in the United States following introduction of the virus [20]. Both outbreaks occurred in south Texas and both were caused by dengue 1 virus. Fortunately, they were limited in time, geographic distribution and magnitude. These two occurrences, after an absence of 35 years, underscore the increased potential for epidemic dengue transmission in the United States, as well as in Europe, where a competent mosquito vector also occurs [1, 18].

In the United States, two mosquito vectors are well established, *Ae. aegypti* and *Ae. albopictus*. *Ae. aegypti* originated in Africa and has been in the United States for over 200 years [22]. *Ae. albopictus* is an Asian species and was first detected in 1985 after being imported in used automobile tires [19]. Because *Ae. albopictus* has the capability to diapause, it can withstand cold temperatures. As a result, this species has expanded its range in the United States as far north as Chicago. This species currently has a wide distribution in the eastern United States.

*Ae. albopictus* was also recently introduced into Europe (Italy and Albania) and has expanded its geographic distribution in those countries [1, 18]. Whether it will expand its range to more northern countries in Europe remains to be seen, but the experience in the United States suggests that this is possible. If it does, there is the potential for autochthonous transmission secondary to importation of the virus in travelers in more northern latitudes of Europe. Dengue virus importation into Europe has been repeatedly documented in recent years [3, 16, 17, 26, 28].

The occurrence of severe disease associated with imported dengue cases has been a relatively rare event. Thus, it has generally been recommended that travelers not be concerned about severe disease when visiting dengue-endemic areas. In recent years, however, there have been several incidents that may call for a reassessment of this recommendation. In 1991, two adult travelers from Sweden had onset of a dengue-like illness within 24 h of each other approximately 1 week after returning from holiday in Thailand [28]. Both were hospitalized with thrombocytopenia ($< 100000/mm^3$), hemoconcentration and hemorrhagic manifestations. Both cases met the World Health Organization's case definition for DHF [27] and both were confirmed as dengue infection by serology and by isolation of dengue 1 virus. These are interesting cases for two reasons. First, the near simultaneous onset of illness suggests that both were probably infected by the same mosquito, most likely the result of interrupted feeding. Secondly, the fact that both developed DHF is unusual in itself, but is of even more interest because both patients were experiencing their first (primary) dengue infections.

In 1994, 3/46 (6.5%) of patients with laboratory-confirmed dengue infection imported into the United States had severe disease [4]. These included a 12-year-old, hospitalized with dengue shock syndrome (DSS); an 11-year-old hospitalized, with mild disseminated intravascular coagulation; and a 49-year-old with apparent DSS, although capillary leak was not confirmed. Two of the three patients had secondary dengue infections (the third was unknown); the infecting dengue virus serotype was not known for any of these cases.

In 1995, dengue/DHF is the most important arbovirus disease of humans; it has a global distribution in the tropics comparable to malaria, with an estimated 2.5 billion people living in areas at risk for epidemic transmission (Fig. 3). Each year there are numerous major epidemics in urban centers of the tropics. More than 50 million cases of dengue fever and in excess of 200 000 cases of the severe disease (DHF or DSS) are estimated to occur annually. The case-fatality rate for DHF/DSS in most countries is about 5%, primarily involving children.

The reasons for this dramatic global emergence of dengue and DHF as major public health problems are complex and not well understood [13]. However, several important factors can be identified. First, effective mosquito control is virtually nonexistent in most dengue-endemic countries. The recommended method for the past 20 years has been ultra low volume insecticide space sprays for adult mosquito control, an ineffective approach for *Ae. aegypti* [7, 9, 11]. Second, there have been major global demographic changes in the past two decades. The most important of these is unprecedented population growth with concurrent uncontrolled and unplanned urbanization. This has resulted in inadequate housing and water, sewer, and waste management systems, all of which facilitate transmission of mosquito-borne disease in tropical urban environments. Third, increased human movement by airplane provides the ideal mechanism for transporting dengue viruses between population centers of the tropics, resulting in a constant movement and exchange of dengue viruses and other pathogens. Lastly, the public health infrastructure has deteriorated badly in most countries of the world. Limited financial and human resources and competing priorities have resulted in a crisis mentality, with emphasis on implementing so-called emergency control methods in response to epidemics rather than developing programs designed to prevent epidemic transmission. This approach has been particularly detrimental to dengue control because in most countries surveillance is very poor; the system to detect increased transmission normally relies on reports by local physicians, who often have a low index suspicion for dengue even in endemic countries [9].

As a result, an epidemic has often reached or passed the peak of transmission before it is detected and reported.

Currently, no dengue vaccine is available for disease prevention, but attenuated candidate vaccine viruses have been developed in Thailand [2]. These have been shown to be safe and immunogenic when given in various formulations, including a quadravalent vaccine for all four dengue virus serotypes. However, efficacy trials in human volunteers have yet to be initiated. Research is also being conducted to develop second-generation recombinant vaccine viruses by using the Thailand attenuated viruses as templates. Unfortunately, it will probably be 5 to 10 years before an effective dengue vaccine is available for public use.

Prospects for reversing the recent trend of increased epidemic activity and geographic expansion of dengue are not very good for the near future. Thus, there are constant introductions of new dengue virus strains and serotypes into areas where the population densities of *Ae. aegypti* are the greatest in history. With no new mosquito control technology available, emphasis in recent years has been placed on disease prevention and mosquito control through larval source reduction using community participation [9, 14, 23]. While it is likely that this approach may be effective in the long-term, it is unlikely to impact disease transmission in the near future. It is important, therefore, to develop improved, proactive, laboratory-based surveillance systems that can provide early warning of an impending dengue epidemic [11]. At the very least, surveillance results can alert the public to take action and physicians to diagnose and properly treat patients with dengue or DHF.

Similarly, prospects for reversing the trend of increased epidemic activity of other arboviruses are poor. And with continued frequent travel to tropical destinations by people from temperate areas, increasingly larger numbers of imported dengue and other arboviral diseases can be expected. Moreover, with the occurrence of competent mosquito vectors in many temperate regions of the world, there will be increased potential for outbreaks of autochthonous disease transmission. It is important that public health officials and physicians recognize this threat and implement more effective, laboratory-based surveillance for all arboviral diseases.

## References

1. Adhami J, Murati N (1987) Presence du moustique *Aedes albopictus* en Albanie. Rev Mjekesore 1: 13–16
2. Bhamarapravati N (1993) Dengue vaccine development. Monograph on dengue/dengue haemorrhagic fever, WHO, New Delhi
3. Braito A, Nicoletti L, Pantini C (1994) Dengue di importazione: segnalazione di un caso. G Mal Infet Parassit 46: 334–337
4. Calisher CH, Beaty BJ (1992) Arboviruses. In: Lennette EH (ed) Laboratory diagnosis of viral infections, 2nd ed. Marcel Dekker New York, pp 243–279
5. CDC (1995) Imported dengue – United States, 1993 and 1994. MMWR 44: 353–356

6. U.S. International Air Travel Statistics (1983–94). Department of Transportation, Washington, DC

7. Gratz NG (1991) Emergency control of *Aedes aegypti* as a disease vector in urban areas. Emergency Control of *Ae. agypti* 7: 353–365

8. Gubler DJ (1988) Dengue. Chapter 23, Vol. II, In: Monath TP (ed) Epidemiology of arthropod-borne viral disease, CRC Press, Boca Raton, pp 223–260

9. Gubler DJ (1989) *Aedes aegypti* and *Aedes aegypti*-born disease control in the 1990s: top down or bottom up. Am J Trop Med Hyg 40: 571–578

10. Gubler DJ (1993) Emergent and resurgent arboviral diseases as public health problems. In: Mahy BWJ, Lvov DK (eds) Concepts in virology: from Ivanovsky to the present. Proceedings of Ivanovsky Institute International Symposium, "100 Years of Virology," St. Petersburg, September 21–25, 1992. Harwood City, pp 257–273

11. Gubler DJ, Casta-Valez A (1991) A program for prevention and control of epidemic dengue and dengue hemorrhagic fever in Puerto Rico and the U.S. Virgin Islands. Bull PAHO 25: 237–247

12. Gubler DJ, Sather GE (1990) Laboratory diagnosis of dengue and dengue hemorrhagic fever. Proceedings of International Symposium on Yellow Fever and Dengue, Rio de Janeiro, Brazil, pp 291–322

13. Gubler DJ, Trent DW (1994) Emergence of epidemic dengue/dengue hemorrhagic fever as a public health problem in the Americas. Infect Agents Dis 2: 383–393

14. Halstead SB (1993) Community-based dengue control: a description and critique of the Rockefeller Foundation program. Trop Med 35: 285–291

15. Karabatsos N ed. (1985) International catalogue of arboviruses including certain other viruses of vertebrates, 3rd ed. American Society of Tropical Medicine Hygiene, San Antonio

16. Ligtenberg JJM, Hospers GAP, Sprenger HG, Weits J (1991) Hemorragische koorts door dengue bij twee toeristen. Ned Tijdschr Geneeskd 135: 2394–2397

17. Lopez-Vélez R, Tapia-Ruano C, Garcia-Camacho A, Rodolfo Sanchez R (1994) Dengue: enfermedad importada del subcontinente indio. Enferm Infecc Microbiol Clin 12: 182–186

18. Possa GD, Majori G (1992) First record of *Aedes albopictus* establishment in Italy. JAMCA 8: 318–320

19. Reiter P, Sprenger D (1987) The used tire trade: a mechanism for the worldwide dispersal of container breeding mosquitoes. JAMCA 3: 494–501

20. Rigau-Pérez JG, Gubler DJ, Vorndam AV, Clark GG (1994) Dengue surveillance – United States, 1986–1992. MMWR 43: 7–19

21. Rosen L, Gubler DJ (1974) The use of mosquitoes to detect and propagate dengue viruses. Am J Trop Med Hug 23: 1153–1160

22. Rush B (1789) An account of the bilious remitting fever, as it appeared in Philadelphia in the summer and autumn of the year 1780. In: Medical inquiries and observations. Prichard and Hall, Philadelphia, pp 89–100

23. Slosek J (1986) *Aedes aegypti* mosquitoes in the Americas: a review of their interactions with the human population. Soc Sci Med 23: 249–257

24. Tsai TF (1994) Arboviruses and related zoonotic viruses. Infect Dis 57: 1266–1288

25. Tsai TF (1995) Arboviruses: In: Murray PR, Baron EJ, Pfaller MA, Tenover FC, Yolken RH (eds) Manual of clinical microbiology, 6th ed. American Society of Microbiology, Washington, pp 980–996

26. Vogtlin J, Gyr K (1985) Dengue-Fieber als Importkrankheit in der Schweiz. Schweiz Med Wochenschr 115: 1273–1277

27. WHO (1980) Guide for diagnosis treatment and control of dengue haemorrhagic fever, 2nd ed. Geneva, Switzerland

28. Wittesjo B, Eitrem R, Niklasson B (1993) Dengue fever among Swedish tourists, Scand J Infect Dis 25: 699–704
29. Wittsesjo B, Eitrem R, Niklasson B, Vene S, Mangiafico JA (1995) Japanese encephalitis after a 10-day holiday in Bali. Lancet 345–856

Author's address: Dr. D. J. Gubler, Division of Vector-Borne Infectious Diseases, National Center for Infectious Diseases, Centers for Disease Control and Prevention, Public Health Service, U.S. Department of Health and Human Services, P.O. Box 2087, Fort Collins, CO 80522, U.S.A.

Arch Virol (1996) [Suppl] 11: 33–40

# Arboviruses causing neurological disorders in the central nervous system

## G. Dobler

Max von Pettenkofer-Institute for Hygiene and Medical Microbiology,
Ludwig-Maximilians-Universität, Munich, Federal Republic of Germany

**Summary.** Arthropod-borne viruses are important causes of diseases of the central nervous system. In addition to the tick-borne encephalitis viruses, other arboviruses in Europe are known to cause neurological disorders. Among them are West Nile, California group, Bhanja, Erve, Kemerovo group, Eyach, and Thogoto viruses. The ecologies and epidemiologies of these viruses are presented and their medical importance as travel-related diseases is discussed.

## Introduction

More than 40 arthropod-borne viruses have been isolated in Europe. When countries of the former Soviet Union are included, the total is 60. More than half of these viruses cause human diseases. The best known and probably most important of the European arboviruses are the agents of European tick-borne encephalitis (Central European encephalitis virus and Russian Spring Summer encephalitis virus, family *Flaviviridae*). Several other arboviruses have been implicated in disorders of the central nervous system (CNS) in humans in Europe (Table 1). The European epidemiology and medical importance of West Nile, California group, Bhanja, Erve, Kemerovo group, Eyach, and Thogoto viruses as etiologic agents of CNS disorders in humans are discussed in some detail.

## West Nile virus

West Nile (WNV) virus, originally isolated in Uganda, is a member of the genus *Flavivirus* (family *Flaviviridae*). It has been isolated many times from various mosquito species and from vertebrates (birds and mammals, including humans) [7]. WNV also has been isolated from ticks although the ecological significance of these isolates is unknown. The natural cycle of WNV involves birds and ornithophilic mosquitoes. The geographic distribution of this virus covers large areas of Africa, Asia and parts of Europe (Fig. 1). In Europe WNV was isolated in Southern France, Cyprus, Slovakia, and Hungary [7]. Antibody studies, however, suggest a greater distribution in southern Europe and also in parts of Central Europe. Over its geographical range WNV is transmitted to humans in

**Table 1.** Arboviruses causing neurological disorders in Europe

| Virus | Principal vector | Distribution in Europe |
|---|---|---|
| *Flaviviridae* | | |
| Flavivirus | | |
| Central European Encephalitis | ticks | throughout |
| Russian Spring-Summer Encephalitis | ticks | Russia |
| Louping ill | ticks | Great Britain, Spain |
| West Nile | mosquitoes | Southern Europe, Slovakia, Hungary |
| *Bunyaviridae* | | |
| Bunyavirus | | |
| Tahyna | mosquitoes | throughout |
| Inkoo | mosquitoes | Northern Europe, Russia |
| Snowshoe hare | mosquitoes | Russia |
| Phlebovirus | | |
| Arbia | sandflies | Italy |
| Toscana | sandflies | Italy, Portugal |
| Nairovirus | | |
| Erve | ticks | France |
| (unplaced) | | |
| Bhanja | ticks | Southern Europe |
| *Reoviridae* | | |
| Orbivirus | | |
| Lipovnik | ticks | Czech Republic, Slovakia |
| Tribec | ticks | Czech Republic, Slovakia, Germany, Russia |
| Coltivirus | | |
| Eyach | ticks | Germany, France |
| *Orthomyxoviridae* | | |
| Thogoto virus | ticks | Southern Europe |

an endemic and, under favourable ecological circumstances, in an epidemic transmission cycle.

WNV infections in humans can manifest with various symptoms. Illness appears suddenly and can be accompanied by fever, malaise, frontal headache, photophobia, myalgia, arthralgia, and lymphadenopathy, some or all manifestations lasting about 5 to 7 days [7]. A maculopapular rash occurs in about half of the cases. Gastrointestinal disorders, respiratory symptoms and more severe clinical manifestations, including meningitis, meningo-encephalitis, myocarditis, pancreatitis, and hepatitis have been reported [7]. These severe and sometimes fatal clinical forms of WNV infections tend to occur predominantly in patients of older ages.

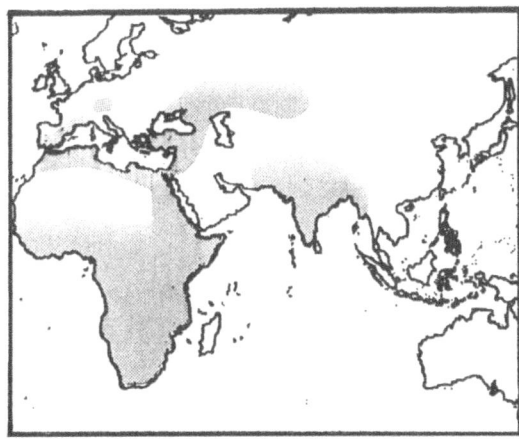

**Fig. 1.** Geographical distribution of West Nile virus

The importance of WNV as a travel-related disease is unclear. The high endemicity, especially in parts of the mediterranean region (Egypt, Israel), suggests that travellers could come in contact with the virus. The usually mild clinical course, a "flu-like" disease in most cases probably will not lead the patient to visit a physician. It is likely that most cases are misdiagnosed as dengue fever (serological cross-reaction), malaria or other infectious diseases.

### California encephalitis viruses

At least three viruses of the California (CAL) group (genus *Bunyavirus*, family *Bunyaviridae*), Tahyna, Inkoo and snowshoe hare viruses are known to occur in Europe [9]. Tahyna virus has been isolated over large areas of central and eastern Europe and in Asia (Fig. 2). Inkoo virus is mainly found in northern Europe and northeastern Russia (Fig. 2). Snowshoe hare virus occurs in western Russia. In some of these areas, two and possibly three CAL group viruses co-circulate (Fig. 2). Under these circumstances, it is possible for genome reassortment to occur between these viruses.

The natural cycles of CAL group viruses in Europe involve mosquitoes of various species (*Aedes* species and *Culex* species) as vectors. Primary vertebrate hosts are hares and rabbits as well as other small- to medium-size vertebrates.

Although CAL group viruses frequently produce inapparent infections in humans, the pathogenic potential of the European CAL group viruses has been well established for several decades. In a Czechoslovakian study the respiratory form of Tahyna virus infection was prominent in humans in about 70% of clinically apparent infections [15], musculoskeletal and abdominal ("appendicitis-like") manifestation forms were observed in about 10% of cases, and involvement of the CNS was observed in about 3% [15].

The primary manifestation of Inkoo virus seems to be a febrile illness with very rare CNS manifestations. Snowshoe hare virus is known to cause febrile illness with frequent CNS involvement. A study in the former U.S.S.R. showed

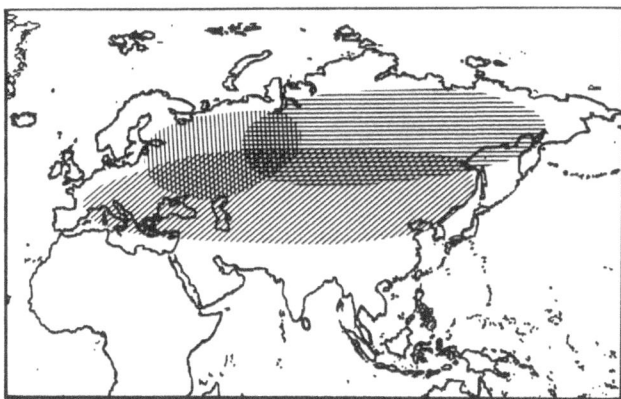

**Fig. 2.** Geographical distribution of California encephalitis group viruses in Eurasia. ☐ Tahyna virus; ▨ Inkoo virus; ▨ Snowshoe hare virus

a frequency of respiratory involvement of 65% in CAL group virus infections [10]. In the same study CNS involvement of CAL group virus infections in humans was reported to occur at a rate of 35%. Unfortunately the study did not differentiate between the CAL serotypes. Possibly most of the severe cases were caused by snowshoe hare virus or by reassortment CAL viruses.

The importance of CAL group viruses as travel-related diseases is unknown. Prevalence and incidence studies of travellers have not been done so far. Snowshoe hare virus, the virus that appears to have the greatest pathogenicity for humans, is found in sparsely inhabited regions. However, the importance of the northern tundra regions of Eurasia is increasing, because of its wealth of natural resources. Increasing numbers of technologists, miners, and also tourists are travelling to these regions. One study showed high antibody prevalences to several CAL group viruses in U.S. Geological Service and U.S. Forest Service workers in Alaska [16]. In this study seropositivity strongly correlated with travel to remote areas in Alaska. Unpublished allusions to hundreds of thousands of cases of CAL group virus infections with CNS involvement have been reported from essentially throughout Russia within the past few years (CH Calisher, pers. comm., 1994). Whether these infections are associated with CAL group virus infections or only with antibody to these viruses has not been determined. CAL group viruses, possibly natural reassortant mutants, may yet be incriminated in these infections.

### Bhanja virus

Bhanja virus (family *Bunyaviridae*) has not been placed in any of the four genera of the family [9]. This virus was first isolated from ticks in India, but subsequent studies showed that it is distributed over large geographical areas in Africa, Asia, and Europe (Fig. 3). Virus isolations have been reported from ixodid ticks, cattle, sheep, goats and humans (natural and laboratory infections with illnesses). The natural cycle of this virus is not clear. Likely, Bhanja virus occurs in econiches, that include ixodid ticks as vectors and domestic animals (sheep or cattle) and, possibly small mammals, as vertebrate hosts.

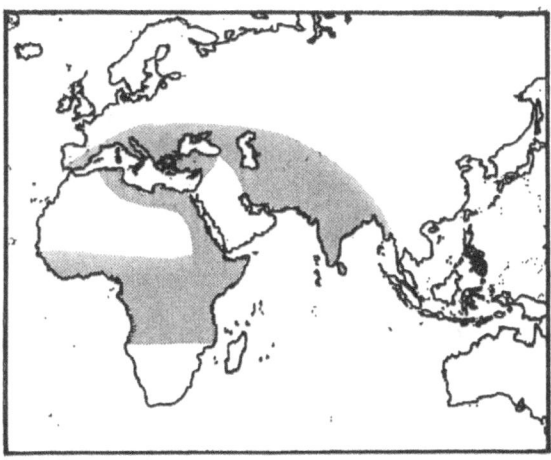

**Fig. 3.** Geographical distribution of Bhanja virus

Subclinical human infections occur and the clinical picture may be as severe as meningoencephalitis. One subclinical laboratory infection and one laboratory infection with mild symptoms (fever, weakness, photophobia, frontal headache) have been described [13]. One naturally acquired Bhanja virus infection was reported, with symptoms of a severe meningoencephalitis, including back pain, fever, photophobia, vomiting, meningeal signs, hyperreflexia, and unconsciousness for several days [17].

Because of the ecological specificity of the natural cycle of this virus, Bhanja virus infections are not a major threat to travellers. However, travelling in endemic areas with intensive contact with domestic animals can increase the risk of infection with this virus.

### Erve virus

Erve virus was isolated from shrews in France in 1982 [2]. Subsequent studies showed that this virus is a member of the genus *Nairovirus*, family *Bunyaviridae* [18]. Serosurveys indicate that Erve virus has a wide distribution, at least throughout France. Antibodies against Erve virus were found in rodents, insectivores, large forest animals (wild boars, deers), and humans [2].

There is evidence that Erve virus can cause disease in humans. One child exhibiting a fatal hemorrhagic syndrome showed complement fixing (CF) antibodies against Erve virus [2]. Three other patients with neurological disorders, and exhibiting CF antibodies against Erve virus have so far been reported [2]. Further studies must show whether the geographical distribution of Erve virus is restricted to France, and what relevance this virus has as a cause of human illness.

### Kemerovo group viruses

The viruses of the Kemerovo group form a distinct serogroup in the genus *Orbivirus*, family *Reoviridae*. Six Kemerovo group viruses have been isolated in

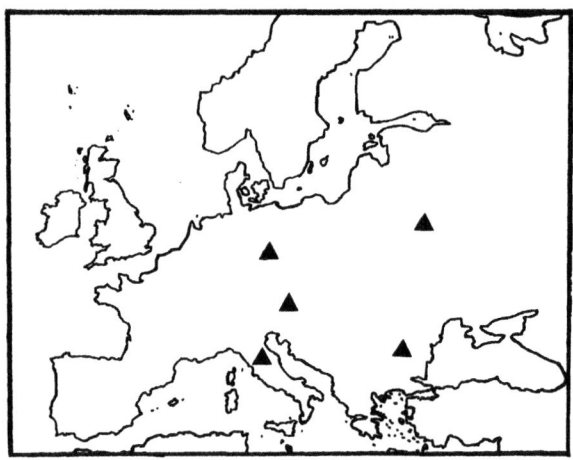

**Fig. 4.** Isolations in Europe of viruses of the genus *Orbivirus* with pathogenic potential for humans

Europe [10]. Of these, Tribec and Lipovnik viruses have been incriminated as causes of human illness. Both viruses are transmitted by ixodid ticks, but the complete natural cycle of these viruses is unknown. Kemerovo virus has been isolated from small rodents and ticks [9] and serologic evidence of infection has been found in domestic cattle and humans. Isolations of Tribec and Lipovnik viruses have been reported from Slovakia, Italy, Romania, Belorussia, and Germany (Fig. 4) [9].

Viruses of the Kemerovo complex were first associated with human illness, when Kemerovo virus was isolated from cerebrospinal fluid of patients with meningoencephalitis in West Siberia in 1962 [3]. Since then, on several occasions Kemerovo group viruses were found as agents of meningo-encephalitis in the former U.S.S.R. [4]. In Slovakia, antibodies to Tribec virus were found in cerebrospinal fluid of patients with inflammatory neurologic disorders (M. Labuda, pers. comm., 1992). Viruses of the Kemerovo complex were incriminated in aggravating the clinical symptoms of tick-borne encephalitis in concurrent infections [9]. Detailed data on the distribution and medical importance of Kemerovo group viruses in Europe are lacking.

**Eyach virus**

Eyach virus was isolated in 1972 from *Ixodes* (*Ix.*) *ricinus* ticks in southern Germany [14]. A second isolation was reported from *Ix. ventalloi* and *Ix.ricinus* ticks in Brittany, France [1]. Subsequently, Eyach virus was found to be serologically related to Colorado tick fever virus. Studies showed that these two viruses were distinct from other tick-borne orbiviruses. For these reasons, they recently were reclassified into the newly created genus *Coltivirus* within the familiy *Reoviridae* [12].

Eyach virus so far is the only virus related to Colorado tick fever virus found outside of Northern America. Limited studies did not show antibodies against

Eyach virus in any of the tested animals. Therefore, the origin and the natural cycle of this virus remain obscure.

The close serologic relationship of Eyach and Colorado tick fever viruses implies a possible role of Eyach virus in pathogenicity in human disease. IgM antibodies were found in the cerebrospinal fluids of patients with meningoencephalitis, meningoradiculitis and polyneuritis in Czechoslovakia [5, 11]. The role of Eyach virus in the pathogenicity of human neurological disorders remains to be elucidated.

## Thogoto virus

Thogoto virus originally was isolated from ticks in Kenya [6]. Subsequently other isolations of this virus were reported, mainly from ticks and domestic animals in sub-Saharan Africa [9]. Subsequently, Thogoto virus was isolated from ixodid ticks in Ethiopia, Egypt, and Sicily, which expands the distribution of Thogoto virus to the Mediterranean [9]. Thogoto virus is classified in the family *Orthomyxoviridae*; its genus placement is uncertain. Domestic animals (cattle and camels) appear to play a possible role in the natural cycle of the virus.

Human infections with Thogoto virus are characterized by fever and CNS involvement; encephalitis, meningo-encephalitis were reported. Other clinical symptoms observed during infection with Thogoto virus include optic neuritis. Human cases described so far are few, therefore the medical importance of Thogoto virus infections in Africa and especially in the Mediterranean remains unclear.

Bhanja, Erve, Kemerovo, Eyach, and Thogoto viruses do not seem to play an important role as travel-related diseases. However, physicians should be aware of them. An arboviral etiology should be ruled out in neurological disorders of unknown origin in patients returning from areas where these viruses are endemic.

## Acknowledgement

The author wishes to thank Charles H. Calisher for critical review and help in the preparation of the manuscript.

## References

1. Chastel C, Main AJ, Couatarmanac HA, LeLay G, Knudson DL, Quillien MC, Beaucournu JC (1984) Isolation of Eyach virus (Reoviridae, Colorado Tick fever group) from *Ixodes ricinus* and *Ixodes ventalloi* ticks in France. Arch Virol 82: 161–167
2. Chastel C, Main AJ, Richard P, LeLay G, Quillien MC, Beaucournu JC (1989) Erve virus, a probable member of Bunyaviridae family isolated from shrews (*Crocidura russula*) in France. Acta Virol 33: 270–280
3. Chumakov MP, Karpovich LG, Sarmanova ES (1963) Report on the isolation from *Ix. persulcatus* ticks and from patients in Western Siberia of a virus differing from the agent of tick-borne encephalitis. Acta Virol 7: 82
4. Chunikhin SP, Alekseev AN (1989) The status and main directions of arbovirology development in the USSR. In: Uren MF, Blok J, Manderson LH (eds) Arbovirus

research in Australia. Commonwealth Scientific and Industrial Research Organisation, Brisbane, pp 358–362

5. Frankova V (1981) Meningoencephalitis caused by arboviruses of the genus *Orbivirus* in C.S.S.R. Shorn Lek 83: 234–235

6. Haig DA, Woodall JP, Danskin DG (1965) Thogoto virus: a hitherto undescribed agent isolated from ticks in Kenya. J Gen Microbiol 38: 389–394

7. Hayes CG (1989) West Nile fever. In: Monath TP (ed) The arboviruses: epidemiology and ecology vol V. CRC Press, Boca Raton, pp 59–88

8. Hubalek Z (1986) Comparative biogeography of Bhanja virus in Europe. Acta Sci Nat Brno 20: 1–50

9. Karabatsos N (ed) (1985) International catalogue of arboviruses, 3rd ed. American Society of Tropical Medicine and Hygiene, San Antonio

10. Kolobukhina LV, Lvov DK, Butenko AM, Nedyalkova MS, Kuznetsov AA, Galkina IV (1990) Signs and symptoms of infections caused by California serogroup viruses in humans in the USSR. In: Calisher CH (ed) Hemorrhagic fever with renal syndrome, tick- and mosquito-borne viruses. Springer, Wien New York, pp 243–247 (Arch Virol [Suppl] 1)

11. Malkova D, Holubova J, Kolmar JM, Marhoul Z, Hanzal F, Markvart H, Simkova L (1980) Antibodies against some arboviruses in persons with various neuropathies. Acta Virol 24: 298

12. Murphy FA, Fauquet CN, Bishop DHL, Ghabrial SA, Jarvis AW, Martelli GP, Mayo MA, Summers MD (eds) (1995) Virus Taxonomy. Classification and Nomenclature of Viruses. Sixth Report of the International Committee on Taxonomy of Viruses. Springer, Wien New York (Arch Virol [Suppl] 10)

13. Punda V, Beus I, Calisher CH, Vesenjak-Hirjan J (1980) Laboratory infections with Bhanja virus. In: Vesenjak-Hirjan J, Porterfield JS, Arslanagic E (eds) Arboviruses in the mediterranean countries. Zbl Bakteriol [Suppl] 9: 273–275

14. Rhese-Küpper B, Casals J, Rhese E, Ackermann R (1976) Eyach – an arbovirus related to Colorado tick fever virus in the Federal Republic of Germany. Acta Virol 20: 339–346

15. Sluka M (1969) The clinical picture of Tahyna virus infection. In: Blaskovic D (ed) Arboviruses of the California virus complex and the Bunyamwera group. Slovak Academy of Sciences Press, Bratislava, pp 311–314

16. Stansfield SK, Calisher CH, Hunt AR, Winkler WG (1988) Antibodies to arboviruses in an Alaskan population at occupational risk of infection. Can J Microbiol 34: 1213–1216

17. Vesenjak-Hirjan J, Calisher CH, Beus I, Marton E (1980) First natural clinical human Bhanja virus infection. In: Vesenjak-Hirjan J, Porterfield JS, Arslanagic E (eds) Arboviruses in the mediterranean countries. Zbl Bakteriol [Suppl] 9: 297–301

18. Zeller HG, Karabatsos N, Calisher CH, Digoutte JP, Cropp CB, Murphy FA, Shope RE (1989) Electron microscopic and antigenic studies of uncharacterized viruses. II. Evidence suggesting the placement of viruses in the family *Bunyaviridae*. Arch Virol 108: 211–227

Author's address: Dr. G. Dobler, Max von Pettenkofer-Institute for Hygiene and Medical Microbiology, Ludwig-Maximilians-Universität, Pettenkoferstrasse 9a, D-80336 München, Federal Republic of Germany.

Arch Virol (1996) [Suppl] 11: 41–47

# Sandfly fever viruses in Italy

**L. Nicoletti, M. G. Ciufolini,** and **P. Verani**

Laboratory of Virology, Istituto Superiore di Sanità, Rome, Italy

**Summary.** Two serologically distinct agents, the sandfly fever Sicilian and the sandfly fever Naples viruses, were isolated by Sabin from blood samples taken during an Italian epidemic of febrile illness. Since then, several different viruses have been isolated from sandflies and/or humans in both the Old and New World. Toscana virus, a new virus closely antigenically related to sandfly fever Naples virus, was isolated in 1971 from the sandfly *Phlebotomus perniciosus* in Italy. Extensive studies on the importance of Toscana virus as a human pathogen demonstrated its association with acute neurologic diseases. A serosurvey for the presence of antibodies to sandfly fever Sicilian, sandfly fever Naples and Toscana viruses indicated that, as in other Mediterranean areas, both sandfly fever Sicilian and sandfly fever Naples viral infections decreased or disappeared after the 1940s in countries performing insecticide-spraying malaria eradication campaigns. In contrast, clinical cases of aseptic meningitis or meningoencephalitis caused by Toscana virus are observed annually in Central Italy during the summer. Toscana virus may be present in other Mediterranean countries where sandflies of the genus *Phlebotomus* are present.

## Historical review

As Sabin has pointed out in his review of the history of phlebotomus fever, Pick first described sandfly fever disease in 1886 [25]. At the same time, the disease was already known in Italy as "Pappataci fever", suggesting a possible link with sandflies. Since then, sandfly or phlebotomus fever has been a disease of considerable medical importance, mainly for the military. Typically, outbreaks of sandfly fever occur when large numbers of nonimmune individuals, such as military personnel, enter an endemic area, as was the case during World War II for Allied Troops in Italy in 1943–44 [23] as well as for German, British and American troops stationed in North Africa and in the Mediterranean region, and for British colonial troops in India and Pakistan [2, 15]. Two serologically distinct agents, the sandfly fever Sicilian and the sandfly fever Naples viruses, were isolated by Sabin from blood samples taken during the Italian epidemic [24]. Since then several different viruses have been isolated from sandflies and/or humans in both the Old and New World and, on the basis of their serologic reactivity, these viruses have been classified in the phlebotomus fever group of arboviruses [16, 36, 37]. Some of them have been suspected of causing human infection, although in some cases, only on the basis of the presence of antibodies [30]. For seven different viruses, including sandfly fever Sicilian and Naples, overt disease has been described [16, 29].

During field studies, in 1971, on the ecology of arboviruses, a new virus closely antigenically related to sandfly fever Naples virus, was isolated in Italy from the sandfly *Phlebotomus perniciosus* and was named Toscana after the region of Central Italy where the sandflies were collected. In addition, antibodies in healthy individuals were relatively highly prevalent in the region of isolation, indicating that the virus infects man [20, 42]. Extensive studies on the importance of Toscana virus as a human pathogen demonstrated its association with acute neurologic diseases [19] and this is the only sandfly-transmitted phlebovirus with such neurotropic activity.

## Clinical presentation

Phlebotomus fever is a mild, self-resolving, flu-like illness. The incubation period ranges from 2–6 days. Fever is always present ranging from 38 to 41 °C; it has an abrupt onset, with a duration from a few hours to four days (phlebotomus fever is also known as "three day fever"). Myalgia and headache are often observed. In some cases low back pain, retro-orbital pain, conjunctivitis, photophobia and malaise, nausea, vomiting, dizziness and neck stiffness are present. Occasionally, the illness resembles early meningitis [14]. Viremia is quite transient (24 to 36 h). Patients completely recover within 1 to 2 weeks. No deaths associated with phlebotomus fever have been reported [1, 4, 10, 13, 14, 23, 27, 29].

IgM antibodies, present in serum of patients 4–5 days after the onset of symptoms, reach their highest titer after 1 to 4 weeks, and persist for at least 8 months. High titers of neutralizing antibodies are present in convalescent sera.

## Acute neurologic disease

An extensive survey conducted in different regions of Central Italy identified more than 500 cases of meningitis or meningoencephalitis associated with Toscana virus infection [19]. The disease has an incubation of at least five days and begins as a classical phlebotomus fever, with fever and headache 2 to 4 days before the appearance of more serious symptoms, elevated fever ($>38$ °C), myalgia, neck rigidity and a positive Kernig's sign. Mental confusion and lethargy can also occur. The hematologic picture is similar to phlebotomus fever. In many cases, there is increased cerebrospinal fluid pressure ($>200$ mm/Hg). Pleocytosis is always present, ranging from 11 to 1 400 cells/mm$^3$. The protein concentration is elevated, from 0.4 to 1.8 mg% [22]. Clinical abnormalities usually resolve in a few days, and patients recover spontaneously within 1–2 weeks, with severe and persistent headache as the only recorded sequela [6, 9]. In one patient, the initial picture of the disease mimicked an Herpes simplex virus type 1 infection [6].

IgM antibodies, usually present at the onset of symptoms, can persist for more than one year. IgG antibodies can be absent at the onset of symptoms; their titers rise in convalescent sera and persist for many years, probably conferring long-lasting immunity. By ELISA test, both IgM and IgG antibodies were found in cerebrospinal fluid [6, 19]. As patients live in areas where Toscana virus is endemic, periodic reinfections could also be responsible for persistent antibodies. By radioimmunoprecipitation test, the main immunologic response of both IgM and IgG subclasses is directed against the nucleoprotein N. Antibodies against this protein are usually present at the onset of symptoms and their titers rise in convalescent sera; IgG against N protein can be detected for at least two years after recovery (Nicoletti et al., unpubl. data).

The virus has been isolated repeatedly from CSF but not from blood, even when samples were collected at the same time, suggesting that the virus is already cleared from the blood at the onset of neurologic symptoms.

## Geographic distribution, incidence, and prevalence

Phlebotomus fever viruses have been isolated from sandflies in Southern Europe, Africa, Central Asia and the Americas and there is evidence for the presence of different viruses in the same sandfly population. Tesh et al. [35] documented the simultaneous occurrence of sandfly fever Sicilian, Karimabad and Isfahan (a *Vesiculovirus*) viruses in Iran, and Verani et al. [41] isolated Toscana (associated with human disease) and Arbia (not yet shown to cause infection in man) viruses in Italy in the same sandfly population. Sandfly fever Naples and Sicilian viruses have the widest geographical distribution, in parallel to their vector's (*P. papatasi*) distribution. Reports of isolations and serologic studies indicate that sandfly fever Naples and Sicilian viruses are present in the Mediterranean coastal regions of Europe and North Africa, the Nile valley, most of South West Asia, areas adjacent to the Black and Caspian Seas, and in Central Asia including Bangladesh [38].

The Old World phleboviruses, such as sandfly fever Naples, sandfly fever Sicilian, Karimabad and Toscana, are relatively ubiquitous and are known to cause epidemics of human infection and/or disease. The sandflies carrying these viruses, *P. papatasi*, *P. perfiliewi* and *P. perniciosus*, come closely in contact with man, readily entering houses to feed. In some rural communities of endemic areas in the Old World, the prevalence of phlebovirus infection is quite high among the indigenous human population. For example, Tesh et al. [38] found prevalences as high as 62% for Karimabad virus in some provinces of Iran, and in Egypt Saidi et al. [26] reported prevalences of 59% and 56% for sandfly fever Sicilian and Naples viruses, respectively.

Antibodies to sandfly fever Sicilian, sandfly fever Naples and Toscana viruses surveyed in 1977 in Italy [20] indicated that, as in other Mediterranean areas, both sandfly fever Sicilian and sandfly fever Naples viral infections decreased or disappeared after the 1940s in countries performing insecticide-spraying malaria eradication campaigns [35]. In fact, antibodies were detected only in people born before the malaria control program began, which was probably the result of a dramatic reduction of the *P. papatasi* population to levels not permitting virus transmission. In contrast, age-specific antibody rates suggest that Toscana virus, which is transmitted by *P. perniciosus* and *P. perfiliewi* with breeding habits different from *P. papatasi*, is still endemic in Italy. A high infection rate (24.8%) was observed among residents of the region where the virus was first isolated [20].

Indirect evidence of the presence of Naples, Sicilian and Toscana viruses in Cyprus has been obtained during serological studies on Swedish troops and Swedish tourists travelling in that country. [10, 11, 21].

Clinical cases of aseptic meningitis or meningoencephalitis caused by Toscana virus are observed every year in Central Italy during the summer, with peaks in August, when the sandflies vector are most active, and when virus strains are isolated from insects collected in natural foci [19, 41]. The incidence of cases of meningitis due to Toscana virus is linked to climatic conditions which influence the annual density of the vector sandflies.

Toscana virus infections can produce only mild symptoms, and it is possible that only severe cases, involving CNS disease and requiring hospitalization, are recognised. Sporadic cases of CNS disease due to infection with Toscana virus have been reported; one occurred in a Swede visiting Portugal [9], and others in an American and a German returning from central Italy, where natural foci of Toscana virus have been identified [5, 28]. These reports indicated that Toscana virus may be present in other Mediterranean countries where sandflies of the genus *Phlebotomus* are known to occur [3].

## Transmission cycles

Many studies on the maintenance of sandfly-transmitted viruses in nature have shown that the main mechanism seems to be transovarial transmission in this vector, a hypothesis which has been confirmed by isolation of phleboviruses from naturally-infected, non-hematophagous male sandflies [1, 31, 39, 41]. Laboratory studies have confirmed that some phleboviruses can be transmitted transovarially from both parenterally and orally infected females to their progeny [7, 8, 12, 17, 34], and Toscana and Arbia viruses have been maintained in laboratory colonies of *P. perniciosus* by vertical transmission for several consecutive generations [7]. Nevertheless, it has been observed that the virus infection rates in each subsequent generation of the colony gradually decreased [34], suggesting that phleboviruses cannot be maintained in the vector population by transovarial transmission alone, and that alternative mechanisms of amplification may occur in nature. Laboratory studies demonstrated venereal infection of of *P. perniciosus* females mated to males transovarially infected with Toscana virus and this may be an accessory amplification mechanism for some phleboviruses in nature [7, 32]. A third possible maintenance mechanism, which would further enhance virus survival, is the vertebrate-insect cycle. However, although some phleboviruses have been isolated from the blood of sick persons and from wild animals [5, 16, 20, 40, 41], the importance of vertebrates in the maintenance cycle of these agents in unclear. Viremia is present after phlebovirus infection in humans and in susceptible laboratory animals only for short periods and at low titer [4, 18]. Moreover, experimental studies have shown that sandflies are relatively refractory to oral infection with a number of phleboviruses. A large quantity of virus must be ingested in order to infect the flies [8, 33], so it seems unlikely that a biting sandfly would often come in contact with an animal with a viremia high enough to infect the insect orally. These observations support the hypothesis that, in order for the virus to survive, amplification of a vertebrate-insect cycle occurs only occasionally in nature.

## References

1. Aitken THG, Woodall JP, De Andrade AHP, Bensabath G, Shope RE (1975) Pacui virus, phlebotomine flies, and small mammals in Brazil: an epidemiological study. Am J Trop Med Hyg 24: 358–368

2. Anderson WME (1941) Clinical observations on sandfly fever in the Peshawar district. J R Army Med Corps 77: 225

3. Balducci M, Fausto AM, Verani P, Caciolli S, Renzi A, Paci P, Amaducci L, Leoncini F, Volpi G (1985) Phlebotomus-transmitted viruses in Europe. In: Proceedings of the International Congress for Infectious Diseases. Luigi Pozzi, Rome, pp 101–104

4. Bartelloni PJ, Tesh RB (1976) Clinical responses of volunteers infected with phlebotomus fever group virus (Sicilian type). Am J Trop Med Hyg 25: 456–462

5. Calisher CH, McLean RG, Smith GC, Szmyd DM, Muth DJ, Lazuick JS (1977) Rio Grande – a new phlebotomus fever group virus from south Texas. Am J Trop Med Hyg 26: 997–1004

6. Calisher CH, Winberg AN, Muth DJ, Laznick JS (1987) Toscana virus infection in a United States citizen returning from Italy. Lancet i: 165–166

7. Ciufolini MG, Maroli M, Guandalini E, Marchi A, Verani P (1989) Experimental studies on the maintenance of Toscana and Arbia viruses (Bunyaviridae: *Phlebovirus*). Am J Trop Med Hyg 40: 669–675

8. Ciufolini MG, Maroli M, Verani P (1985) Growth of two Phleboviruses after experimental infection of their suspected sandfly vector, *Phlebotomus perniciosus* (Diptera: Psychodidae). Am J Trop Med Hyg 34: 174

9. Ehrnst A, Peters CJ, Niklasson B, Svedmyr A, Holmgren B (1985) Neurovirulent Toscana virus (a sandfly fever virus) in Swedish man after a visit to Portugal. Lancet i: 1212–1213

10. Eitrem R, Niklasson B, Weiland O (1991) Sandfly fever among Swedish tourists. Scand J Infect Dis 23: 451–457

11. Eitrem R, Vene S, Niklasson B (1990) Incidence of sandfly fever among Swedish United Nations soldiers on Cyprus during 1985. Am J Trop Med Hyg 43: 207–211

12. Endris RG, Tesh RB, Young DG (1983) Transovarial transmission of Rio Grande virus (Bunyaviridae: *Phlebovirus*) by the sandfly, *Lutzomyia anthophora*. Am J Trop Med Hyg 32: 862–864

13. Feinsod FM, Ksiazek TG, Scott RMcN, Soliman AK, Farrag IH, Ennis WH, Peters CJ, El Said S, Darwish MA (1987) Sandfly-fever infection in Egypt. Am J Trop Med Hyg 37: 193–196

14. Fleming J, Bignall JR, Blades AN (1947) Sandfly-fever. Review of 664 cases. Lancet i: 443–445

15. Hertig M, Sabin AB (1964) Sandfly fever. In: Coates JB (ed) Preventive Medicine in World War II, vol 7. Communicable diseases. US Government Printing Office, Washington, DC, pp 109–174

16. Karabatsos N (ed) (1985) International catalogue of arboviruses including certain other viruses of vertebrates. American Society of Tropical Medicine and Hygiene, San Antonio

17. Maroli M, Ciufolini MG, Verani P (1993) Vertical transmission of Toscana virus in the sandfly *Phlebotomus perniciosus* via the second gonotrophic cycle. Med Vet Entomol 7: 283–286

18. McLean RG, Szmyd DM, Calisher CH (1992) Experimental studies of Rio Grande virus in rodent hosts. Am J Trop Med Hyg 31: 569–573

19. Nicoletti L, Verani P, Caciolli S, Ciufolini MG, Renzi A, Bartolozzi D, Paci P, Leoncini F, Padovani P, Traini E, Baldereschi M, Balducci M (1991) Central nervous system involvement during infection by Phlebovirus Toscana of residents in natural foci in Central Italy (1977–1988). Am J Trop Med Hyg 45: 429–434

20. Nicoletti L, Verani P, Lopes MC, Ciufolini MG, Zampetti P (1980) Studies on Phlebotomus-transmitted viruses in Italy. II. Serological status of human beings. Zentralbl Bakteriol [Suppl] 9: 203–208

21. Niklasson B, Eitrem S (1985) Sandfly fever among Swedish UN troops in Cyprus. Lancet i: 1212

22. Paci P, Balducci M, Verani P, Coluzzi M, Amaducci L, Leoncini F, Nicoletti L, Ciufolini MG, Fratiglioni L (1983) Toscana virus, a new *Phlebotomus*-transmitted virus isolated in Italy. In: Proceedings of the International Congress for Infectious Disease. Luigi Pozzi, Rome, pp 35–39

23. Sabin AB, Philip CB, Paul JR (1944). Phlebotomus (pappataci or sandfly) fever: a disease of military importance; summary of existing knowledge and preliminary report of original investigations. JAMA 125: 603–606, 693–699

24. Sabin AB (1951) Experimental studies on phlebotomus (pappataci, sandfly) fever during World War II. Arch Virusforsch 4: 367–410

25. Sabin AB (1959) Phlebotomus fever. In: Rivers TM, Horsfall FL (eds) Viral and rickettsial infections of man, 3rd ed. Lippincott, Philadelphia, pp 454–460

26. Saidi S, Tesh RB, Javadian E, Sahabi Z, Nadim A (1977) Studies on the epidemiology of sandfly fever in Iran. II. The prevalence of human and animal infection with five Phlebotomus fever serotypes in Isfahan province. Am J Trop Med Hyg 26: 288–293

27. Sandler A (1946) The clinical picture of pappataci fever, especially in Palestine. Med J Aust 1: 789

28. Schwarz TF, Gilch S, Jäger G (1993) Travel-related Toscana virus infection. Lancet 342: 803–804

29. Srihongse S, Johnson CM (1974) Human infections with Chagres virus in Panama. Am J Trop Med Hyg 23: 690–693

30. Tesh RB (1988) The genus *Phlebovirus* and its vectors. Annu Rev Entomol 33: 169–181

31. Tesh RB, Chaniotis BN, Peralta PH, Johnson KM (1974) Ecology of viruses isolates from panamamian sandflies. Am J Trop Med Hyg 24: 258–266

32. Tesh RB, Lubroth J, Guzman H (1992) Simulation of arbovirus overwintering: survival of Toscana virus (Bunyaviridae: *Phlebovirus*) in its natural sandfly vector *Phlebotomus perniciosus*. Am J Trop Med Hyg 47: 574–581

33. Tesh RB, Modi GB (1987) Maintenance of Toscana virus in *Phlebotomus perniciosus* by vertical transmission. Am J Trop Med Hyg 36: 189–193

34. Tesh RB, Modi GB (1984) Studies on the biology of *Phleboviruses* in sandflies (Diptera: *Psychodidae*) I. Experimental infection of the vector. Am J Trop Med Hyg 33: 1007–1016

35. Tesh RB, Papaevangelou G (1977) Effect of insecticide spraying for malaria control on the incidence of sandfly fever in Athens, Greece. Am J Trop Med Hyg 26: 163–166

36. Tesh RB, Peralta PH, Shope RE, Chaniotis BN, Johnson KM (1975) Antigenic relationship among Phlebotomus fever group arboviruses and their implications for the epidemiology of sandfly fever. Am J Trop Med Hyg 24: 135–144

37. Tesh RB, Peters CJ, Meegan JM (1982) Studies on antigenic relationship among *Phleboviruses*. Am J Trop Med Hyg 31: 149–155

38. Tesh RB, Saidi S, Gajdamovic S Ja, Rodhain F, Vesenjak-Hirjan J (1976) Serologic studies on the epidemiology of sandfly fever in Old World. Bull World Health Organ 54: 663–674

39. Tesh RB, Saidi S, Javadian E, Nadim A (1977) Studies on the epidemiology of sandfly fever in Iran. I. Virus isolates obtained from *Phlebotomus*. Am J Trop Med Hyg 26: 282–287

40. Travassos da Rosa APA, Tesh RB, Pinheiro FP, Travassos de Rosa JFS, Peterson NE (1983) Characterization of eight new Phlebotomus fever serogroup arboviruses (Bunyaviridae: *Phlebovirus*) from the Amazon region of Brazil. Am J Trop Med Hyg 32: 1164–1171

41. Verani P, Ciufolini MG, Caciolli S, Renzi A, Nicoletti L, Sabatinelli G, Bartolozzi D, Volpi G, Amaducci L, Coluzzi M, Paci P, Balducci M (1988) Ecology of viruses isolated from sandflies in Italy and characterization of a new *Phlebovirus* (Arbia virus). Am J Trop Med Hyg 38: 433–439

42. Verani P, Lopes MC, Nicoletti L, Balducci M (1980) Studies on Phlebotomus- transmitted viruses in Italy. I. Isolation and characterization of a sandfly fever Naples-like virus. Zentralbl Bakeriol [Suppl] 9: 195–201

Authors' address: Dr. L. Nicoletti, Laboratory of Virology, Istituto Superiore di Sanità, Viale Regina Elena 299, I-00161 Rome, Italy.

Arch Virol (1996) [Suppl] 11: 49–55

# Vector-borne viral diseases in Sweden – a short review

B. Niklasson[1, 2] and S. Vene[1]

[1] Swedish Institute for Infectious Disease Control, Stockholm,
[2] National Defense Research Establishment, Umeå, Sweden

**Summary.** Ockelbo disease, caused by a Sindbis-related virus transmitted to man by mosquitoes, was first described in the central part of Sweden in the 1960s as clusters of patients with fever, arthralgia and rash. An average annual rate of 30 cases was recorded in the 1980s but no cases have been diagnosed during the last few years. Nephropathia epidemica (NE) characterized by fever, abdominal pain and renal dysfunction has been known to cause considerable morbidity in Sweden during the last 60 years but the etiologic agent (Puumala virus) was not isolated until 1983. This virus's main reservoir is the bank vole (*Clethrionomys glareolus*). NE is endemic in the northern two thirds of Sweden where more than a hundred cases are diagnosed each year. Tick-borne encephalitis transmitted by *Ixodes ricinus* ticks is restricted to the archipelago and Lake Mälaren on the east coast close to Stockholm. Between 30 and 110 cases are diagnosed every year. Inkoo virus, a California encephalitis group virus, has been isolated from mosquitoes in Sweden. The antibody prevalence to Inkoo virus is very high in a normal population, but no disease has as yet been associated with this virus in Sweden. Among the vector-borne virus diseases imported to Sweden, dengue is the most important, with approximately 50 cases recorded every year.

## Ockelbo disease

In the latter half of the 1960s, clusters of cases with fever, arthralgia, and rash were observed in Sweden. The disease, called Ockelbo disease, was caused by infection with a Sindbis-like virus [27], and epidemics involving a similar disease were described later in Finland as Pogosta disease, and in Russia as Karelian fever [1, 2, 15, 16, 23]. In the 1980s, Ockelbo disease caused considerable human morbidity in areas of northern Europe with outbreaks involving hundreds of cases. Russia reported 200 and Finland 300 laboratory-confirmed cases in 1981 [1, 2, 15, 16, 23]. In Sweden, an annual average of 30 laboratory-confirmed cases were recorded during the 1980s [13]. The number decreased during the last 4 years as shown in Table 1. Most diagnosed individuals are of working age (most between 30 and 60 years of age) with an equal sex distribution [13]. Age-adjusted antibody prevalence of 4% was measured by plaque reduction neutralization (PRNT) assay in the endemic area [13]. Cases occur mainly between the 60th and 64th parallels, and start to appear in the late July with a peak in August with the last cases seen in October [13]. This period coincides with the season of mushroom and berry picking, a very popular activity in this part of the world. It has not yet been determined whether the peak incidence in mid-August is due to the extent of viral dissemination in nature, increased human exposure, or both.

Both field and experimental laboratory studies have documented that Ockelbo virus, like other Sindbis and Sindbis-like viruses, is maintained in nature in an enzootic bird-mosquito cycle. Ockelbo virus has been isolated from *Culiseta (Cs) spp.*, *Cs. morsitans*, *Culex (Cx) pipiens/torrentium* and *Aedes (Ae.) cinerius* [5, 19]. Transmission experiments found *Cx. torrentium*, *Ae. cinerius*, *Ae. communis* and *Ae. excrucians* to be highly susceptible and competent vectors whereas *Cx. pipiens* was a poor transmitter [11, 29].

The antigenic variation found between strains from different geographic regions is most often "one-way", where the strain is indistinguishable in a one directional test, but can be differentiated in a complete cross PRNT [14]. Strains from Northern Europe (Sweden and Russia) are indistinguishable by PRNT. Sequence analysis has demonstrated that strains from northern Europe are more closely related to South and Central African strains than to the Egyptian prototype, and that Indian and Australian strains belong to another distinct branch of the evolutionary tree [35]. One theory is, therefore, that Ockelbo virus was introduced recently by migrating birds from southern or central parts of Africa.

## Nephropathia epidemica

An illness characterized by sudden onset of fever, headache, severe abdominal pain, and renal dysfunction was first described in Sweden in 1934 by Myhrman and Zetterholm [17, 36]. The name Nephropathia epidemica (NE), suggested by Myhrman, is now generally accepted. Myhrman was an internist and the title of his first publication on NE was "A new renal disease with peculiar symptomatology", while Zetterholm, who was a surgeon, wrote a paper entitled "Acute nephritis that can mimic acute abdomen". NE still causes differential diagnostic problems, and patients may be subjected to acute laparatomy because of a severe abdominal status.

It later became clear that diseases clinically identical or closely related to NE occurred in large parts of Europe and Asia. The name hemorrhagic fever with renal syndrome (HFRS), including both NE and Korean hemorrhagic fever (KHF), was suggested by Gajdusek in 1962 and is now widely accepted, although hemorrhagic manifestations are noted only in a minority of the patients [6].

During 1985–1991, the average annual number of NE cases in Sweden was 153. The number of serologically confirmed cases 1991–1994 is seen in Table 1. The population in the endemic area of Sweden is 2.3 million, which gives an average annual incidence of 7 (range 0.4–12) per 100 000 inhabitants in that area

**Table 1.** Vector-borne viruses in Sweden (serologically verified cases)

| Year | SFF | Dengue | JE | TBE | Ockelbo | NE |
|------|-----|--------|-----|-----|---------|-----|
| 1991 | 5 | 49 | 0 | 68 | 5 | 289 |
| 1992 | 2 | 62 | 0 | 75 | 9 | 184 |
| 1993 | 0 | 50 | 0 | 46 | 0 | 186 |
| 1994 | 0 | 46 | 1 | 116 | 0 | 113 |

[18] and the male:female ratio is approximately 2:1. Age-adjusted antibody prevalence is 8% in the 3 most endemic counties of northern Sweden [24]. Comparing antibody prevalence with the number of diagnosed cases suggests the ratio of cases to infections at approximately 1:14–1:20 in the male and female populations, respectively [22]. The etiological agent of NE, Puumala (PUU) virus, was isolated in 1983 from a bank vole (*Clethrionomys glareolus*) in Sweden [18, 26, 34] and a nationwide study of NE in Sweden identified the major endemic area of the disease, and showed that *C. glareolus* is an important vector of PUU virus [24]. *C. glareolus* is also the most abundant small mammal in Sweden and has a range from the southern tip of the country close to the Arctic circle. Although only one subspecies of *C. glareolus* is recognized throughout the Swedish mainland, a gradient exists in population stability. Populations of voles in southernmost Sweden are non-cyclic, in middle Sweden the populations are intermediate, whereas populations in the north fluctuate on a three- to four-year cycle of abundance [7, 9]. During peak abundance, voles may be >300 times more abundant than immediately after population declines in northern Sweden [9]. There is a boundary between the cyclic populations in the north and the intermediate or non-cyclic populations in the south, and this boundary corresponds to the *Limes Norrlandicus*, a bio-geographical border running from 59 °N on the west coast of Sweden to 61 °N on the east coast. The *Limes Norrlandicus* also appears to demarcate the southern distribution boundary of NE in humans and of PUU virus in voles. A recent serological study among voles trapped in the endemic area provided further evidence that *C. glareolus* is the most important vector of PUU virus. Both the abundance and the antibody prevalence were higher for this species than for *C. rufocanus* and *Microtus agrestis* [25].

## Tick-borne encephalitis

Tick-borne encephalitis (TBE) is typically a biphasic disease. The initial stage (1–8 days) includes symptoms such as fever, headache and myalgia. After an asymptomatic interval of a few days to 3 weeks, patients develop meningo-encephalitis [37]. The first case of TBE in Sweden was described in 1954, and TBE virus was isolated in 1958 from a patient and from *Ixodes ricinus* [32, 28]. Most patients are infected in the archipelagos and at the coastline of the Baltic sea and Lake Mälaren close to Stockholm [8]. The endemic area has been stable over the last three decades when cases have been recorded. Between 1956 and 1989, an average of 30–50 cases per year were diagnosed with a fatality rate of less than 1% [8], and the cases recorded 1991–1994 are listed in Table 1. Most cases occur in July-September. The median age of TBE patients is 40 years and the male:female ratio is approximately 2:1.

## Other vector-borne viruses in Sweden

Batai virus and Inkoo viruses have been isolated from mosquitos in Sweden [5]. A total of 882 sera from normal population collected at 8 different locations distributed from northern to southern Sweden were tested for presence of antibodies to Batai and Inkoo viruses by PRNT. No sera were positive to Batai virus while 549 (62%) of the sera had a PRNT titer of 1:10 or more to Inkoo virus. The Inkoo virus antibody prevalence showed a gradient with a low antibody prevalence in the south (Dalby, 8%), medium in central Sweden (38%–53%), and high antibody prevalence figures in northern Sweden (Umeå, 90% and Kiruna, 80%) (Niklasson, unpubl. obs.). Acute and convalescent sera from 46 patients with encephalitis and negative for TBE, enterovirus and herpes virus, were also negative for Inkoo virus antibodies (Niklasson and M. Forsgren, unpubl. obs.).

## Vector-borne viral diseases imported to Sweden

Dengue fever is the single, most common imported vector-borne viral disease in Sweden. Of 161 patients diagnosed in 1991–1993, most (86%) contracted the infection during travel in Southeast Asia (53% visiting Thailand) [31]. The number of diagnosed patients during 1991–1994 can be seen in Table 1. The first case of Japanese encephalitis in a short term tourist visiting Bali was diagnosed in 1994 [33].

In 1984, seven clinical cases of sandfly fever (SE) were confirmed serologically in Swedish UN troops stationed in Cyprus [20] and one was diagnosed in a Swedish tourist visiting the island in the same season. In 1986–1989, a total of 37 cases of SF occurred among Swedish patients. One patient was diagnosed with a Naples virus infection, one with Toscana virus, and the remaining 35 with Sicilian virus infections. Except for the patient with a Toscana virus infection, who had visited Spain [3] all patients were short term tourists visiting the same resort in Cyprus. The epidemiological data was sent to the local health authority on Cyprus and appropriate vector control measures were initiated. There have been no diagnosed cases during the last 2 years (Table 1).

After travelling in sub-Saharan Africa, an area where sporadic cases of Marburg virus infection are known to occur, a young Swedish man presented in 1991 with a classical picture of severe viral hemorrhagic fever, complicated by disseminated intravascular coagulation and septicemia [10]. Serum samples examined by electron microscopy revealed particles of a size compatible with filovirions. Indirect fluorescent antibody tests indicated a transient seroconversion to Marburg virus. In lymphocyte transformation assays of cells isolated from the patient 11 months after onset of acute disease, Marburg viral antigen stimulated lymphocyte proliferation 3.9-fold; however, exhaustive attempts to isolate virus from acute blood cultured in vitro or in vivo in guinea-pigs and monkeys failed. Data suggest that this patient may have been infected with a filovirus but no definitive diagnosis was made. This case demonstrates not only the difficulties that may occur in laboratory diagnosis of viral hemorrhagic fevers [10], but also the difficulties facing the nursing staff of an intensive care ward [4].

# References

1. Brummer-Korvenkontio M, Kuusisto P (1981) Has western Finland been spared the "Pogosta"? (in Finnish). Suomen Lääkärilehti 36: 2606–2607
2. Brummer-Korvenkontio M, Kuusisto P (1984) Pogosta disease. In: Lvov DK (ed) Proc. Int. Symp. Etiology, Epidemiology, Diagnosis and Prophylaxis of Karelian Fever-Pogosta Disease. Academy of Medical Sciences, Moscow
3. Eitrem R, Niklasson B, Weiland O (1991) Sandfly fever among Swedish tourists. Scand J Infect Dis 23: 451–457
4. Foberg U, Frydén A, Isaksson B, Jahrling P, Johnson A, McKee K, Niklasson B, Normann B, Peters CJ, Bengtsson M (1991) Viral hemorrhagic fever in Sweden. Experiences from management of a case. Scand J Infect Dis 23: 143–152
5. Francy DB, Jaenson TG, Lundström JO, Schildt EB, Espmark A, Henriksson B, Niklasson B (1989) Ecologic studies of mosquitoes and birds as hosts of Ockelbo virus in Sweden and isolation of Inkoo and Batai viruses from mosquitoes. Am J Trop Med Hyg 41: 355–363
6. Gajdusek DC (1962) Virus hemorrhagic fevers: special reference to hemorrhagic fevers with renal syndrome (epidemic hemorrhagic fever). J Pediatr 60: 841–857
7. Hansson L, Henttonen H (1985) Gradient in density variations of small rodents: the importance of latitude and snow cover. Oecologia 67: 394–402
8. Holmgren B, Forsgren M (1990) Epidemiology of tick-borne encephalitis in Sweden 1956–1989: a study of 1116 cases. Scand J Infect Dis 22: 287–295
9. Hörnfeldt B (1994) Delayed density dependence as a determinant of vole cycles. Ecology 75: 791–806
10. Kenyon RH, Niklasson B, Jahrling PB, Geisbert T, Svensson L, Frydén A, Bengtsson M, Foberg U, Peters CJ (1994) Virological investigations of a case of suspected hemorrhagic fever. Res Virol 145: 397–406
11. Lundström JO, Niklasson B, Francy DB (1990) Swedish *Culex torrentium* and *Cx. pipiens* (Diptera: Culicidae) as experimental vector of Ockelbo virus. J Med Entomol 27: 561–563
12. Lundström JO, Turell MJ, Niklasson B (1990) Effect of environmental temperature of the vector competence of *Culex pipiens* and *Culex torrentium* for Ockelbo virus. Am J Trop Med Hyg 43: 534–542
13. Lundström JO, Vene S, Espmark A, Engvall M, Niklasson B (1991) Geographical and temporal distribution of Ockelbo disease in Sweden. Epidemiol Infect 106: 567–574
14. Lundström JO, Vene S, Saluzzo JF, Niklasson B (1993) Antigenic variation among strains of Sindbis and Sindbis-like viruses as determined by cross neutralization. Am J Trop Med Hyg 49: 531–537
15. Lvov DK, Skvortsova TM, Kondrashina NG, Vershinsky A, Lesnikov AL, Dereviansky VS, Berezina LK, Gromashevsky VL, Adrianova DP, Yakolev VI (1982) Etiology of Karelian fever, a new arbovirus infection Vopr Virusol 6: 690–692 [in Russian]

16. Lvov DK, Berezina LK, Yakolev BI, Aristova VA, Gushchina EL, Lvov SD, Myas-nikova IA, Skvortsova TM, Gromashevsky VL, Gushchin BV, Sidorova GA, Klimenko SM, Khutoretskaya NV, Khizhnyakova TM (1984) Isolation of Karelian fever agent from *Aedes communis* mosquitoes. Lancet ii: 399
17. Myhrman G (1934) En njursjukdom med egenartad symptombild. Nord Med Tidskr 7: 793–794
18. Niklasson B, LeDuc J (1984) Isolation of the nephropathia epidemica agent in Sweden. Lancet i: 1012–1013
19. Niklasson B, Espmark A, LeDuc JW, Gargan TP, Ennis WA, Tesh RB, Main AJ (1984) Association of a Sindbis-like virus with Ockelbo disease in Sweden. Am J Trop Med Hyg 33: 1212–1217
20. Niklasson B, Eitrem R (1985) Sandfly fever among Swedish UN troops in Cyprus. Lancet i: 1212
21. Niklasson B, LeDuc J (1987) Epidemiology of nephropathia epidemica in Sweden. J Infect Dis 155: 269–276
22. Niklasson B, LeDue J, Nyström K, Nyman L (1987) Nephropathia epidemica: incidence of clinical cases and antibody prevalence in an endemic area. Epidemiol Infect 99: 559–562
23. Niklasson B (1988) Sindbis and Sindbis-like viruses. In: Monath TP (ed) The arboviruses: epidemiology and ecology, vol 4. CRC Press, Boca Raton, pp 167–176
24. Niklasson B, Hörnfeldt B, Mullaart M, Settergren B, Tkachenko E, Myasnikov YuA, Ryltceva EV, Leschinskaya E, Malkin A, Dzagurova T (1993) An epidemiological study of hemorrhagic fever with renal syndrome in the Bashkiria (Russia) and Sweden. Am J Trop Med Hyg 48: 670–675
25. Niklasson B, Hörnfeldt B, Lundkvist A, Björsten S, LeDuc J (1995) Temporal dynamics of Puumala virus antibody prevalence in voles and of nephropathia epidemica incidence in humans. Am J Trop Med Hyg 53: 134–140
26. Schmaljohn CS, Hasty SE, Dalrymple JM, LeDuc JW, Lee HW, von Bonsdorff CH, Brummer-Korvenkontio M, Vaheri A, Tsai TF, Regnery HL, Goldgaber D, Lee PW (1985) Antigenic and genetic properties of viruses linked to hemorrhagic fever with renal syndrome. Science 227: 1041–1044
27. Skogh M, Espmark A (1982) Ockelbo disease: epidemic arthritis-exanthema syndrome in Sweden caused by Sindbis virus-like agent. Lancet i: 795–796
28. Svedmyr A, von Zeipel G, Holmgren B, Lindahl J (1958) Tick-borne meningoence-phalomyelitis in Sweden. Arch Ges Virusforsch 8: 565–576
29. Turell MJ, Lundström JO (1990) Effect of environmental temperature on the vector competence of *Aedes aegypti* and *Aedes taeniorhynchus* for Ockelbo virus. Am J Trop Med Hyg 43: 543–550
30. Turell MJ, Lundström JO, Niklasson B (1990) Transmission of Ockelbo virus with *Aedes cinerius*, *Aedes communis*, and *Aedes excrusians* (Diptera: Culicidae) collected in an enzootic area in central Sweden. J Med Entomol 27: 226–228
31. Vene S, Mangiafico J, Niklasson B (1994) Indirect immunofluorescence for serological diagnosis of dengue infections in Swedish patients. Clin Diagn Virol 4: 43–50
32. von Zeipel G (1959) Isolation of viruses of the Russian spring and summer encephalitis–louping ill group from Swedish ticks and from a human case of meningo-encephalitis. Arch Ges Virusforsch 9: 460–469
33. Wittesjö B, Eitrem R, Niklasson B, Vene S, Mangiafico J (1995) Japanese encephalitis after a ten-day holiday in Bali. Lancet i: 856
34. Yanagihara R, Goldgaber D, Lee PW, Amyx HL, Gajdusek DC, Gibbs Jr CJ, Svedmyr A (1984) Propagation of Nephropathia epidemica agent in cell culture. Lancet i: 1013

35. Yukio S, Niklasson B, Dalrymple JM, Strauss GE, Strauss JH (1991) Structure of the Ockelbo virus genome and its relationship to other Sindbis viruses. Virology 182: 753–764
36. Zetterholm SG (1934) Akuta nefriter simulerande akuta bukfall. Sv Läkartidningen 31: 425–429
37. Ziebart-Schroth A (1972) Frühsommermeningoencephalitis (FSME). Klinik und besondere Verlaufsformen. Wien Klin Wochenschr 84: 778–781

Authors' address: Dr. B. Niklasson, Swedish Institute for Infectious Disease Control, S-105 21 Stockholm. Sweden.

Arch Virol (1996) [Suppl] 11: 57–65

# Travel-related vector-borne virus infections in Germany

**T. F. Schwarz**[1], **G. Jäger**[1], **S. Gilch**[1], **C. Pauli, M. Eisenhut**[1],
**H. Nitschko**[1], and **B. Hegenscheid**[2]

[1] Max von Pettenkofer Institute for Hygiene and Medical Microbiology,
Ludwig Maximilians University, Munich, [2] Institute of Tropical Medicine,
Berlin, Federal Republic of Germany

**Summary.** Laboratory diagnosis of imported, vector-borne virus diseases during a 22-month-period in Munich, Germany, is summarized. In 13/317 Germans returning from the Mediterranean with suspected sandfly fever, acute sandfly fever, serotype Toscana, was confirmed serologically: 84.6% of the infections were acquired in Italy. Of 249 German tourists with febrile disease returning from the tropics, acute infection with dengue virus was diagnosed serologically in 26 (10.4%): most infections were acquired in Thailand (57.7%). In a seroepidemiological study of 670 German aid workers who had spent two years in the tropics, 49 (7.3%) were positive for antibodies to dengue, 9 (1.3%) to chikungunya, and 1 (0.1%) to Sindbis virus. Of 17 Middle Eastern patients with suspected viral haemorrhagic fever, genomic Crimean-Congo haemorrhagic fever virus RNA was amplified in 4 (23.5%) by semi-nested reverse transcriptase polymerase chain reaction, and confirmed by molecular characterization of nucleic acid. With the increase in travel to and from endemic areas, imported vector-borne virus infections are increasingly important in Germany.

## Introduction

Until the 1960s, public health concern about imported virus diseases in Germany mainly centered on the importation of variola virus. In 1967, a previously unknown virus (Marburg virus) caused an outbreak among laboratory staff handling newly imported monkeys from Uganda [30]. The 1980s and 1990s were characterized by the emergence of several "newly" identified vector- and rodent-borne viral pathogens. Of these, Ebola Reston, Guanarito, Sabiá, Puumala virus, Dobrova virus, sin nombre virus and other related hantaviruses, are the most significant [22]. Some of these viruses are endemic in areas visited often by persons from non-endemic countries, and therefore may pose a potential risk of infection.

Although numerous cases of vector-borne infections associated with travelling were reported in recent years, the true incidence is unknown. Most frequently, dengue virus infection was reported in travelers returning from southeast

Asia, the Caribbean, and Oceania [6, 7, 10, 19, 21, 25, 29, 33, 34]. The virus is transmitted by *Aedes aegypti* and *Aedes albopictus*, and may cause dengue fever, dengue haemorrhagic fever, or dengue shock syndrome [9, 16]. In most travelers infected in these areas, dengue fever was observed, and only few cases of dengue haemorrhagic fever were reported [9, 33, 34]. The widespread re-emergence of dengue virus in recent years [16] means that dengue must be considered the most frequently imported vector-borne virus infection.

Sandfly fever virus, serotype Toscana (TOSV), transmitted by *Phlebotomus (P.) perniciousus* and *P. perfiliewi* causes a febrile illness with headaches, photophobia, myalgia and arthralgia, sometimes complicated by aseptic meningitis [23]. Sandfly fever (pappataci fever) was medically significant in the Mediterranean during World War II, requiring hospitalization of 19 000 soldiers [24]. Recent studies have shown that TOSV is still present in some Mediterranean countries and, since the 1980s, several TOSV infections in travelers indicate a growing significance of this reemerging disease [5, 11–13, 26–28].

Crimean-Congo haemorrhagic fever virus (C-CHFV) occurs endemically in eastern Europe, Asia, and Africa [32]. The virus is transmitted by *Hyalomma* ticks. Several nosocomial outbreaks with C-CHFV have occurred in recent years [2, 31]. In the past 20 years, C-CHFV outbreaks were reported from South Africa, the United Arab Emirates, Iraq, Kuwait, Pakistan, Bulgaria, and the former USSR [2–4, 14, 24, 31]. To date, C-CHFV has not been reported as a travel-related infection.

Chikungunya virus (CHIK) is endemic in Africa and Asia, causing a febrile illness associated with severe polyarthritis [1]. Minor haemorrhagic symptoms may occur [32]. The virus is transmitted by *Aedes* and *Mansonia* species. CHIK has rarely been reported as a travel-related infection [17, 33].

Sindbis virus (SIN), which causes an endemic-epidemic mosquito-borne febrile disease, is widely disseminated in Africa, Asia, Australia, the Middle East, and northeastern Europe. Epidemics were reported from South Africa [20]. The virus is transmitted by *Culex* species. Sindbis fever is characterized by fever, exanthema, and arthritis; haemorrhagic manifestations may occur occasionally [15]. At present, it is unknown whether SIN is of any significance as a travel-related infection.

This paper describes vector-borne virus infections diagnosed in Munich, Germany, during a 22-month-period.

## Materials and methods

### *Patients with suspected acute sandfly fever*

From May 1993 to March 1995, we received 328 sera of 317 patients with suspected sandfly fever from various clinics or private practitioners. Of 7 patients with sandfly fever, one or more follow-up sera were available. All 328 sera were examined for antibodies to TOSV (anti-TOSV IgM and IgG) by indirect immunofluorescence assay (IFA) at a dilution of 1:16 (IgM) and 1:32 (IgG). Positive sera were titrated in serial dilutions to determine the end-point. For confirmation, positive sera were retested for anti-TOSV IgM and IgG by a recently developed enzyme-immunoassay (EIA) [27]. Additionally, sera positive for TOSV markers were tested for anti-TOSV IgM, IgA and IgG by a newly developed immunoblot (unpublished) at a dilution of 1:50.

### Patients with suspected dengue fever

Sera of 249 patients with suspected dengue fever were received from May 1993 to March 1995 from various clinics in Germany to test for dengue virus-specific antibodies. All sera were from German tourists returning from endemic countries. Sera were screened for anti-dengue virus IgM and IgG at dilutions of 1:16 (IgM) and 1:32 (IgG). Positive sera were then retested in serial dilutions to determine the end-point.

### Aid workers in Third World countries

A total of 670 individuals who had worked on aid projects in Africa, Asia and Oceania for two years were tested for antibodies to dengue virus, CHIK, and SIN by IFA after returning to Germany. All sera were screened for anti-dengue virus IgG at a dilution of 1:32, and if positive, retested to determine the end-point titer. For detecting anti-CHIK IgG and anti-SIN IgG, sera were also screened at a dilution of 1:32. Sera positive were retested in serial dilutions to determine the end-point.

### Patients with suspected viral haemorrhagic fever

Sera of 17 patients with suspected viral haemorrhagic fever were obtained from various clinics in the Middle East. Infection with C-CHFV was suspected, because the medical history included tick bites or contact with tick-infested animals. For detection of anti-C-CHFV IgM and IgG by IFA, sera were screened for IgM and IgG to C-CHFV at a dilution of 1:16.

### Detection of genomic C-CHFV RNA by semi-nested RT-PCR

Total RNA in sera of the 17 patients with suspected C-CHFV was recovered by addition of an equal volume of guanidine thiocyanate solution, followed by phenol-chlorofom extraction and purification with an RNA matrix (RNaid Kit, Bio 101, La Jolla, CA, USA). For semi-nested RT-PCR, oligonucleotide primers designed from a published sequence of S segment of the C-CHFV genome, were used. All primer sequences were kindly supplied by J. Smith, United States Army Research Institute for Infectious Diseases, Fort Detrick, Frederick, MD, USA. Reverse transcriptase was added to the PCR mix, and the reaction run with 1 cycle of 45 °C for 45 min, then 40 cycles of 94 °C for 40 sec, 38 °C for 40 sec and 72 °C for 1 min 30 sec on a GeneAmp PCR System 9600 thermocycler (Perkin-Elmer). The nested reaction was performed in 40 cycles of 94 °C for 40 sec, 41 °C, for 40 sec and 72 °C for 1 min 30 sec. PCR products were electrophoretically separated on a 2% agarose gel. Nucleic acids of the amplified PCR products were then analysed using the dyedeoxy cycle sequencing method.

### Detection of arbovirus antibodies by indirect immunofluorescence assay

The IFAs used in this study for detecting antibodies to TOSV, dengue virus, C-CHFV, CHIK, and SIN virus were based on the method described for Lassa fever virus [35], and have been described in detail previously [27, 29]. "Spot slides" of Vero-E6 cells infected with the various viruses, and uninfected Vero E6 cells, were mixed, and attached in the same spot.

To detect specific IgM, sera were incubated with RF absorbents (Behringwerke, Marburg, Germany) before being added to the spots.

## Results

### Antibodies to TOS in patients with suspected sandfly fever

Of 317 patients, acute TOSV infection was diagnosed serologically by detecting anti-TOSV IgM in 13 (4.1%). All 11 follow-up sera of 7 of these patients obtained within 3 months after acute infection were also positive for anti-TOSV IgM. Some serological results of these 13 patients with acute infection have been reported previously [26–28]. All 24 sera were positive for anti-TOSV IgM and IgG by EIA. In immunoblot assays using TOSV proteins derived from cultured virus, all sera of patients with acute infection showed a very strong reaction for anti-TOSV IgM, IgA and IgG to the 28 kDa nucleoprotein protein.

All 13 patients with acute sandfly fever were Germans who had vacationed in endemic areas. Medical history of these 13 patients with acute TOSV infection revealed that 11 (84.6%) had been infected in Italy, 1 (7.7%) in Portugal, and 1 (7.7%) in Turkey. Aseptic meningitis of various severity occurred in 10/13 (77.9%) of the patients, and 3 (23.1%) experienced only severe headaches. Of the patients with meningitis, abducens nerve palsy and acute hearing loss occurred in one patient each. No patient suffered residual damage.

### Antibodies to dengue virus in patients with dengue fever

Of 249 sera, dengue fever was confirmed serologically by detecting anti-dengue virus IgM and IgG by IFA and EIA in 26 (10.4%) patients. The results of 17 of these 26 (65.4%) patients have been published previously [29].

All 26 patients with dengue fever were Germans who had vacationed in areas endemic for dengue virus (Table 1). In 20/26 (77.0%) of patients, infection occurred in southeast Asia. Most dengue virus infections (57.7%) were in

Table 1. Imported dengue virus infections in German tourists (n = 26) (May 1993 to March 1995)

| Country | No. of cases (%) |
| --- | --- |
| Thailand | 15 (57.7) |
| Maledives | 2 (7.7) |
| Indonesia | 1 (3.8) |
| Malaysia | 1 (3.8) |
| Laos | 1 (3.8) |
| Venezuela | 3 (11.5) |
| Dominican Republic | 2 (7.7) |
| Ecuador | 1 (3.8) |

**Table 2.** Seroprevalence of antibodies to dengue virus, CHIK, and SIN
in sera (n = 670) of German aid workers tested by IFA

| Country (no. of sera) | dengue pos (%) | CHIK pos (%) | SIN pos (%) |
|---|---|---|---|
| Benin (88) | 13 (14.8) | 5 (5.7) | 0 (0.0) |
| Botswana (32) | 3 (3.1) | 0 (0.0) | 0 (0.0) |
| Burkina Faso (76) | 8 (10.5) | 1 (1.3) | 0 (0.0) |
| Ghana (47) | 2 (4.3) | 0 (0.0) | 0 (0.0) |
| Kamerun (63) | 3 (4.8) | 1 (1.6) | 0 (0.0) |
| Lesotho (46) | 0 (0.0) | 0 (0.0) | 0 (0.0) |
| Papua New Guinea (43) | 3 (7.0) | 0 (0.0) | 0 (0.0) |
| Ruanda (33) | 1 (3.0) | 0 (0.0) | 0 (0.0) |
| Tanzania (65) | 3 (4.6) | 0 (0.0) | 0 (0.0) |
| Thailand (36) | 7 (19.4) | 2 (5.5) | 0 (0.0) |
| Togo (36) | 2 (5.5) | 0 (0.0) | 0 (0.0) |
| Zambia (40) | 1 (2.5) | 0 (0.0) | 1 (2.5) |
| Zimbabwe (65) | 3 (4.6) | 0 (0.0) | 0 (0.0) |

travelers returning from Thailand. Fever, headaches, arthralgia and myalgia were the most frequently noted symptoms in patients with acute dengue virus infection.

### Seroprevalence to arboviruses in German aid workers

The study on the seroprevalence of antibodies to various arboviruses in German aid workers revealed that 49/670 (7.3%) had antibodies to dengue virus, 9/670 (1.3%) to CHIK, and 1/670 (0.1%) to SIN (Table 2). Of these three arboviruses, a relevant antibody prevalence was detected only to dengue virus. For aid workers who had resided in Benin, Burkina Faso, or Thailand, seropositivity rates were 14.8%, 10.5%, and 19.4%, respectively. All sera positive for anti-dengue virus IgG by IFA were confirmed positive by haemagglutination-inhibition assay (kindly performed by H. Holzmann, Vienna, Austria).

### Genomic C-CHFV RNA in patients with viral haemorrhagic fever

Of 17 patients with suspected viral haemorrhagic fever, genomic C-CHFV RNA was amplified by nested RT-PCR in sera of 4 (23.5%) patients giving rise to a 260 bp cDNA fragment. Molecular characterization of the PCR products confirmed the presence of C-CHFV in these 4 sera. All four C-CHFV-RNA-positive patients and the 13 CCHF-RNA-negative patients were negative for antibodies to C-CHFV by IFA (see Fig. 1).

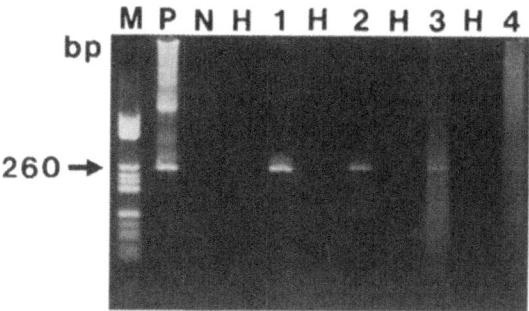

**Fig. 1.** Detection of genomic C-CHFV RNA by semi-nested RT-PCR in serum specimens from the Middle East: marker (*M*), positive control (*P*), negative control (*N*), water controls (*H*), serum specimens *1* to *4*. A 260 bp C-CHFV PCR fragment was amplified from all 4 serum specimens

## Discussion

Infection with sandfly fever virus may occur in travelers returning from Mediterranean countries. We demonstrated here that patients with acute TOSV infection mainly presented with aseptic meningitis after vacationing in Italy, Portugal, and Turkey. The risk of acquiring TOSV infection in endemic Mediterranean areas for travelers from non-endemic countries is unknown, and to date, all reports only describe isolated cases [5, 11–13, 26–28]. In contrast, a large number of TOSV infections with neurological symptoms was reported in an endogenous Italian population [23]. Since a focal occurrence has been reported for TOSV, clusters of infections should be expected in areas where index cases have been observed. Previous studies described the Italian regions of Tuscany and Marche as endemic areas, but recently acute infection was also diagnosed in a German traveler with meningitis on holidays near Naples [28].

Imported dengue virus infections have been reported from several countries. In 1990 and 1991, a total of 49 laboratory-confirmed dengue virus infections were reported in the United States [6, 7]. During 1993 and 1994, a total of 148 cases of imported dengue were suspected among U.S. residents, of which 46 were laboratory-confirmed [9]. In the United Kingdom, 106 cases were reported in 1990 [19]. In Sweden, 106 cases were confirmed serologically from November 1989 to June 1992 [34]. In our study, dengue virus infection was diagnosed in 26 patients from May 1993 to March 1995. In all studies, most imported infections were acquired in Thailand. Since most German tourist destinations, such as Sri Lanka, Maledives, Seychelles, Australia, French Polynesia, Cuba, Dominican Republic, Trinidad and Tobago, Costa Rica, Venezuela, and Brazil, are endemic for dengue viruses, more cases in tourists can be expected.

To date, there are limited data on the occupational risk of arbovirus infections in persons such as aid workers spending longer periods in endemic areas. For military personnel based in endemic areas, arbovirus infections are considered a risk [24]. Dengue virus infections were shown to have caused epidemics among American soldiers in Vietnam, the Philippines and Haiti [8, 18, 24]. Also, infections with CHIK were diagnosed in American soldiers during the Vietnam War [24]. In this study, anti-DEN IgG was found in 7.3% of German aid workers. The highest rates of seropositivity were found in those based in Benin, Burkina Faso, and Thailand. Antibodies to CHIK were less frequent, and were detected mainly in aid workers who had been in Benin and Thailand. Whether CHIK infection is presently of epidemiological and medical importance in these two countries, is unknown. Although most of the aid workers studied here were residing in countries known to be endemic for SIN, only one was antibody-positive, indicating that this infection is not so important as an occupational hazard.

This study describes the detection of genomic C-CHFV RNA by semi-nested RT-PCR in sera of four Middle Eastern patients with haemorrhagic fever. Until now, diagnosis of C-CHFV was based on virus isolation in cell culture and detection of specific C-CHFV antibodies [3, 4]. Working with C-CHFV in the laboratory is hazardous, and requires special containment. Consequently, semi-nested RT-PCR is an improvement for detecting C-CHFV because it is more rapid than virus isolation in cell culture and reduces the risk of laboratory infection. Genomic C-CHFV-RNA was not detected in the sera of the other 13 patients with suspected viral haemorrhagic fever. One explanation may be that these sera had not been stored at $-70\,^{\circ}$C and the shipment to Germany had been delayed. This may have led to the loss of amplifiable viral genomic sequences.

C-CHFV is a rare but often lethal disease. Until now, there has not been any report of travel-related C-CHFV. Since C-CHFV is occurring in areas which are visited by tourists, importation must be considered in the future. Rapid detection of genomic RNA by semi-nested RT-PCR is necessary to confirm or exclude C-CHFV, because the virus may represent a significant risk for nosocomial transmission [2, 14, 31].

Since most physicians in non-endemic countries are unaware of the potential importation of these vector-borne virus diseases and unfamiliar with the symptoms, such infections are rarely suspected. At present laboratory diagnosis of vector-borne virus infections is also hampered by the limited number of institutions performing these tests, but with a growing number of travelers to and from endemic areas, the possibility of these infections has to be drawn into consideration.

## Acknowledgements

Reference virus strains were kindly supplied by the following colleagues: C. Giorgi, Rome, Italy (TOSV), B. Niklasson, Stockholm, Sweden (SFSV, SFNV), A. Antoniades, Thessaloniki, Greece (dengue virus, C-CHFV), and R. Nogueira, Rio de Janeiro, Brazil. Polyclonal antisera to various viruses were supplied by N. Karabatsos, Centers for Disease Control and Prevention, Fort Collins, CO, USA. We are grateful to M. Brand, Munich, for technical assistance.

# References

1. Brighton AW, Prozesky OW, DE La Harpe AD (1983) Chikungunya virus infection. A retrospective study of 107 cases. S Afr Med J 63: 313–315
2. Burney MI, Ghafoor A, Saleen M, Webb PA, Casals J (1976) Nosocomial outbreak of viral hemorrhagic fever caused by Crimean hemorrhagic fever-Congo virus in Pakistan, January 1976. Am J Trop Med Hyg 29: 941–947
3. Burt FJ, Leman PA, Abbott JC, Swanepoel R (1994) Serodiagnosis of Crimean-Congo haemorrhagic fever. Epidemiol Infect 113: 551–562
4. Butenko AM, Chumakov MP (1990) Isolation of Crimean-Congo hemorrhagic fever virus from patients and from autopsy specimens. Arch Virol [Suppl] 1: 295–301
5. Calisher CH, Weinberg AN, Muth DJ, Lazuick JS (1987) Toscana virus infection in United States citizen returning from Italy. Lancet I: 165–166
6. Centers for Disease Control and Prevention (1981) Imported dengue – United States, 1990. Morb Mort Wkly Rep 40: 519–520
7. Centers for Disease Control and Prevention (1993) Imported dengue – United States, 1991. Dengue Surveill Summ 66: 1–4
8. Centers for Disease Control and Prevention (1984) Dengue fever among U.S. military personnel – Haiti, September–November, 1994. Morb Mort Wkly Rep 43: 845–848
9. Centers for Disease Control and Prevention (1995) Imported dengue – United States, 1993–1994. Morb Mort Wkly Rep 44: 353–356
10. Cunningham R, Mutton K (1991) Dengue haemorrhagic fever. Br Med J 302: 1083–1084
11. Ehrnst A, Peters CJ, Niklasson B, Svedmayr A, Holmgren B (1985) Neurovirulent Toscana virus (a sandfly fever virus) in Swedish man after visit to Portugal. Lancet I: 1212–1213
12. Eitrem R, Vene S, Niklasson B (1990) Incidence of sand fly fever among Swedish United Nations soldiers on Cyprus during 1985. Am J Trop Med Hyg 43: 207–211
13. Eitrem R, Niklasson B, Weiland O (1991) Sandfly fever among Swedish tourists. Scand J Infect Dis 23: 451–457
14. Fisher-Hoch SP, Khan JA, Rehman S, Mirza S, Khurshid M, McCormick JB (1995) Crimean Congo-haemorrhagic fever treated with oral ribavirin. Lancet 346: 472–475
15. Guard RW, McAuliffe MJ, Stallman ND, Bramston BA (1982) Haemorrhagic manifestations with Sindbis infection. Case report. Pathology 14: 89–90
16. Gubler DJ, Trent DW (1993) Emergence of epidemic dengue/dengue hemorrhagic fever as a public health problem in the Americas. Infect Agents Dis 2: 383–393
17. Harnett GB, Bucens MR (1990) Isolation of chikungunya virus in Australia. Med J Aust 152: 328–339
18. Hayes CG, O'Rourke TF, Fogelman V, Leavengood DD, Crow G, Albersmeyer MM (1989) Dengue fever in American military personnel in the Philippines: clinical observations on the hospitalized patients during a 1984 epidemic. Southeast Asian J Trop Med Pub Health 20: 1–8
19. Jacobs MG, Brook MG, Weir WRC, Bannister BA (1991) Dengue haemorrhagic fever, a risk of returning home. Br Med J 302: 828–829
20. Jupp PG, Blackburn NK, Thompson DL, Meenehan GM (1986) Sindbis and West Nile virus infections in the Witwatersrand-Pretoria regions. S Afr Med J 70: 218–220
21. Morens DM, Sather GE, Gubler DJ, Rammohan M, Woodall JP (1987) Dengue shock syndrome in an American traveler with primary dengue 3 infection. Am J Trop Med Hyg 36: 424–426
22. Murphy FA, Nathanson N (1994) The emergence of new virus diseases: an overview. Semin Virol 5: 87–102

23. Nicoletti L, Verani P, Caciolli S, Ciufolini MG, Renzi A, Bartolozzi D, Paci P, Leoncini F, Padovani P, Traini E, Baldereschi M, Balducci M (1991) Central nervous system involvement during infection by phlebovirus Toscana of residents in natural foci in central Italy (1977–1988). Am J Trop Med Hyg 45: 429–434

24. Oldfield EC, Wallace MR, Hyams KC, Yousif AA, Lewis DE, Bourgeois AL (1991) Endemic infectious diseases of the Middle East. Rev Infect Dis [Suppl 3] 13: S199–S217

25. Patey O, Ollivaud L, Breuil J, Lafaix C (1993) Unusual neurologic manifestations occurring during dengue fever infection. Am J Trop Med Hyg 48: 793–802

26. Schwarz TF, Gilch S, Jäger G (1993) Travel-related Toscana virus infection. Lancet 342: 803–804

27. Schwarz TF, Jäger G, Gilch S, Pauli C (1995) Serosurvey and laboratory diagnosis of imported sandfly fever virus, serotype Toscana, infection in Germany. Epidemiol Infect 114: 501–510

28. Schwarz TF, Gilch S, Jäger G (1995) Aseptic meningitis caused by sandfly fever virus, serotype Toscana. Clin Infect Dis 21: 669–671

29. Schwarz TF, Jäger G (1995) Imported dengue virus infections in German tourists. Zentral Bakteriol 282: 533–536

30. Siegert R, Shu HL, Slenczka W, Peters D, Müller G (1967) Zur Ätiologie einer unbekannten von Affen ausgegangenen Infektionskrankheit. Dtsch Med Wochenschr 92: 2341–2343

31. Suleiman MNEH, Muscat-Baron JM, Harries JR, Satti AGO, Platt GS, Bowen ETW, Simpson DIH (1980) Congo/Crimean haemorrhagic fever in Dubai – an outbreak at the Rashid hospital. Lancet II: 939–941

32. Swanepoel R (1987) Viral haemorrhagic fevers in South Africa: history and national strategy. S Afr J Sci 83: 80–88

33. Van Tongeren HAE (1981) Imported virus diseases in the Netherlands out of tropical areas 1977–1980 (30 months). Tropenmed Parasit 32: 205

34. Wittesjö B, Eitrem R, Niklasson B (1993) Dengue fever among Swedish tourists. Scand J Infect Dis 25: 699–704

35. Wulff H, Lange JV (1975) Indirect immunofluorescence for the diagnosis of Lassa fever infection. Bull World Health Organ 52: 429–436

Authors' address: Dr. T. F. Schwarz, Stiftung Juliusspital, Central Laboratory, Juliuspromenade 19, D-97070 Würzburg Federal Republic of Germany.

Arch Virol (1996) [Suppl] 11: 67–74

# Imported tropical virus infections in Germany

**H. Schmitz, P. Emmerich,** and **J. ter Meulen**

Department of Virology, Bernhard-Nocht Institute for Tropical Medicine, Hamburg,
Federal Republic of Germany

**Summary.** Our routine tests for tropical viruses document that several hundreds of Dengue fever cases are imported into Germany every year. In contrast, hemorrhagic fever cases are rarely diagnosed in Germany. Our investigations suggest that this low number is due to the different living conditions of the local population in the tropics compared with that of travellers from Europe or North America. Improved methods for detecting Dengue virus infections, e.g. three different antibody tests and the reverse transcriptase-polymerase chain reaction (RT-PCR) for detection of viral RNA, have been developed.

## Introduction

High population density and inadequate sanitary conditions increase the risk of acquiring viral infections in tropical areas, and this holds true for ubiquitous as well as for endemic viral infections. Hepatitis and AIDS are found worldwide but play a dominant role in tropical areas.

In contrast, the classical tropical viral infections are zoonoses, primarily infections of non-human vertebrates (e.g. rodents) and of arthropod vectors, which can be transmitted to man. These rodents and arthropods live in high temperature areas, and are not found in regions of moderate climate.

Due to increased international travel, tropical diseases of humans are no longer confined to tropical areas, but are seen increasingly often by clinicians in Europe or North America. Unfortunately, European clinicians are often unfamiliar with such diseases.

According to the clinical picture, these viral infections can be divided into three basic patterns: a) Influenza-like diseases with arthralgia, b) encephalitis, and (c) hemorrhagic fevers (Table 1). To diagnose the infections caused by the above-mentioned tropical viruses (Table 2), we have prepared numerous tests in our own laboratory.

Most infections diagnosed in our institute belong to the group of influenza like symptoms. Viral hemorrhagic fevers are rare in visitors of tropical areas, which is also reflected by the specimens sent to our institute for diagnosis (Table 2). The most frequent disease is dengue fever whereas all other viruses are only rarely found. Our routine diagnostic data show slightly increasing numbers

**Table 1.** Main clinical symptoms and vectors/animal reservoirs of tropical virus diseases

| Vector/reservoir | Disease | | |
|---|---|---|---|
| | Flu, arthralgia, fever | Encephalitis | Hemorrhagic fever |
| Insects | Dengue<br>Chikungunya<br>Ross river<br>West Nile<br>Sandfly | Japan B<br>St. Louis<br>various equine | Yellow fever<br>Rift valley<br>Crimean-Kongo |
| Rodents | Nephropathia<br>  epidemica | | HFNS<br>Lassa<br>Junin |
| Not known | | | Ebola<br>Marburg<br>Monkey pox |

**Table 2.** Detection of antibodies to tropical viruses during the first quarter of 1995, Institute of Tropical Medicine, Hamburg

| Virus | Suspected cases | Diagnosed cases | Methods |
|---|---|---|---|
| Dengue (Flavi) | 269 | 54 | ELISA, IF, HI |
| Chikungunya | 22 | 2 | HI |
| Japan B | 10 | 1 | ELISA |
| Ross River | 4 | 2 | IF |
| Sandfly | 8 | 0 | IF |
| Rift Valley | 2 | 0 | IF |
| Lassa | 4 | 0 | IF |
| Marburg-Ebola | 3 | 0 | IF |
| HIV-2 | 42 | 2 | WB*, IF |

All tests are home made except the Western blot (WB*). *IF* Indirect immunofluorescence, *HI* hemagglutination inhibition

of dengue fever cases during the last years (Dengue fever patients 1993: 247/712; 1994: 303/1259; confirmed to suspected cases, respectively). It is unclear, however, whether this is due to an increased number of imported cases or to an improved awareness by the clinicians. Compared to Dengue fever, only a few cases were caused by infections with Chikungunya virus or Ross River virus, neither of which can be differentiated from Dengue fever by clinical examination.

Filovirus or Arenavirus infections did not occur at all. During our work in Guinea, however, we diagnosed several Lassa fever cases.

## Epidemiological investigation on the prevalence of Lassa virus in an endemic area in southern Guinea, West Africa

Lassa virus infections are very common in the southern part of the Republic of Guinea. In contrast, infections among European tourists are rarely reported, although many Europeans are travelling in this country. We, therefore, evaluated antibodies to Lassa in southern Guinea, to obtain detailed information on the transmission of Lassa virus from the chronically infected rodent (Mastomys natalensis) to humans. In order to carry out serological investigations locally, we have designed a new enzyme linked immunosorbent assay (ELISA) for anti-Lassa antibodies.

Fragments of the nucleoprotein gene (NP) of the Lassa virus were cloned into the T7-polymerase driven expression vector pJC40 [2], which adds a N-terminal tag of 10 histidines to the recombinant protein. Expression was carried out in *E. coli* BL 21(DE3), transformed with the plasmid pAP2-lacl$^Q$, coding for the lac-repressor, to facilitate expression of toxic proteins. Neither the whole NP nor the N-terminus (aa 1 to 139) could be expressed, but a truncated protein (aa 141 to 569) was abundantly overexpressed and purified by Ni-chelate chromatography. After a process of prolonged renaturation, the protein reacted with human sera positive in immunofluorescence for Lassa antibodies, but not with 5 monoclonals reactive for NP [4]. We conclude that the monoclonals are directed against conformational epitopes that cannot be mimicked by the recombinant protein. Testing a large pannel of positive and negative sera from a Lassa endemic area, and calculating the cut-off of the ELISA as the mean of the negative controls plus three times the SD of the negatives, a sensitivity of only 26.8% but specificity of 98.6% was observed (Fig. 1). For sera with a titre of $\geq 1:80$ the sensitivity was 74% (Fig. 2), indicating that the test might be of valuable for diagnosing acute cases of Lassa fever. This is currently under investigation in Gueckedou, Republic of Guinea, West Africa.

A population-based, randomised epidemiological survey was carried out in Southern Guinea, utilizing the "cluster sampling method", recommended by WHO [3]. A total of 991 persons from 27 villages was enrolled. Lassa antibodies were measured by indirect immunofluorescence and the recombinant Lassa ELISA and correlated with epidemiological data by multivariate logistic regression analysis. The prevalence of Lassa antibodies varied from 2.5% up to 37.5% in selected villages of the endemic area. Both sexes were nearly equally affected and prevalence rates were highest in the age group 40–49 years [5].

All persons reported close contacts with *M. natalensis*, the vector responsible for transmission of the virus. Contact varied from observation of the rodents and their excrements inside the houses and the stored food supplies to hunting and eating of the animals. It appears that specific close contact to *M. natalensis* is the

IF

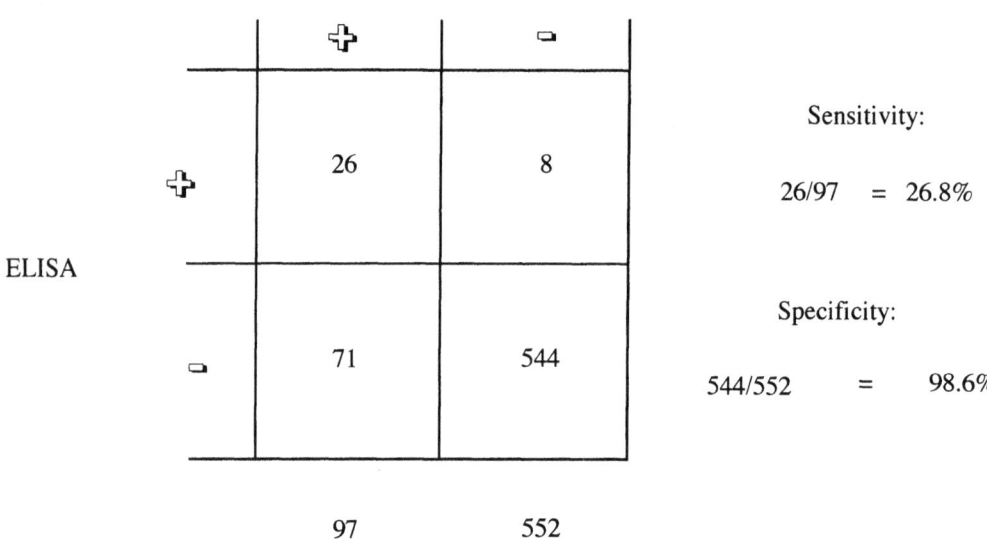

**Fig. 1.** Sensitivity and specificity of the ELISA for the detection of anti-Lassa antibody using the NP antigen produced in *E. coli*. Comparison with the indirect immunofluorescence method

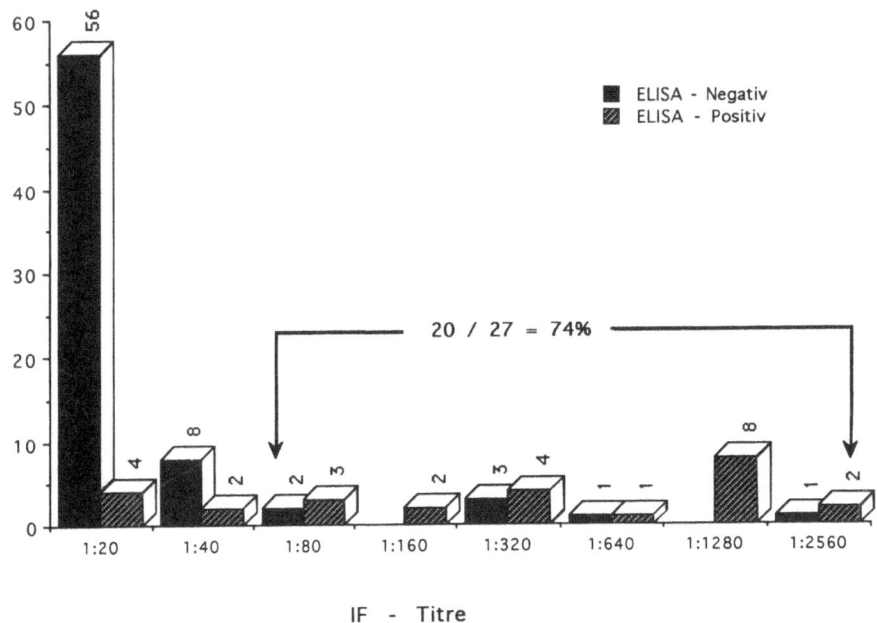

**Fig. 2.** Sensitivity of the ELISA compared to indirect immunofluorescence when only antibody titers of > 1:40 are considered

main risk factor for acquisition of Lassa fever and distinguishes local populations from occasional tourist visitors of the endemic area.

The relative contribution of these risk factors to the prevalence of Lassa antibodies is currently under statistical investigation.

## Ebola and Marburg diagnostics

The Bernhard Nocht Institute is the German Reference Centre for the diagnosis of hemorrhagic fever viruses. With respect to the Ebola or Marburg infections, some guidelines are important for handling suspected cases of Ebola or Marburg fever. As already mentioned for Lassa fever, patients present with acute severe flu-like symptoms (high fever, pharyngitis, diarrhea, vomiting, malaise). Hemorrhagic signs may not be seen early in the course of the disease. Since these patients come from Central Africa, an acute malaria must be excluded. During the first week, virus (both Marburg and Ebola) can already be isolated in Vero cells, and the virus inside the cells can be identified by monoclonal antibodies or by electron microscopy. At the end of the first week after onset of symptoms antibodies (IgM and IgG) can be detected by indirect immunofluorescence. To test for specific anti-Marburg or anti-Ebola virus antibodies, cell smears of infected Vero cells are continuously produced in our class 4 high security facilities. Recently an RT-PCR for Ebola virus genomic RNA has been established which detects Ebola virus RNA in tissue culture supernatant with high sensitivity. However, this method has to be evaluated for its use on clinical specimens.

## Diagnostics of flavivirus infections

In contrast to Lassa fever patients, who were only found in West Africa, we diagnosed an increasing number of imported Dengue fever patients in Germany. Some suffered from severe flu-like symptoms and had elevated aminotransferases and low platelet counts.

The diagnosis of acute Dengue fever was made using three different antibody detection methods. The standard hemagglutination inhibition (HI) test was carried out using mouse brain infected with Dengue type 1 virus as source of glycoprotein antigen. This test has been used for several decades to diagnose acute Flavivirus infections. Due to the various purification and absorption procedures, it is relatively time consuming. In a highly sensitive anti-Flavi IgM test antibodies were bound to a solid phase anti-μ microtiter plate, and the IgM antibodies in turn reacted with viral glycoprotein obtained from tissue culture supernatant (C6/36 cells infected with Dengue type 1 virus). To avoid the purification process necessary for direct labeling of Flavivirus glycoproteins [6], we now detect glycoprotein bound to the patients IgM antibody with a biotin-labeled anti-West Nile monoclonal antibody. This double sandwich technique is completed by adding a Streptavidin-peroxidase conjugate. Using this method, a positive/negative ratio of >20 fold was obtained with human serum specimens. Moreover, all serum samples were also tested for anti-Flavivirus anti-

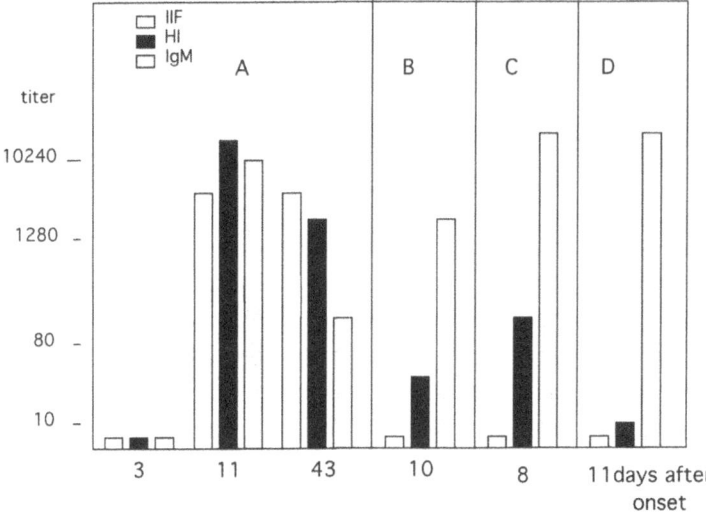

**Fig. 3.** Detection of antibodies to Flavivirues in serum samples obtained from four individuals (*A–D*) during acute Dengue fever using the IgM ELISA, the indirect immunofluorescence (IIF) and the HI test

bodies using the indirect immunofluorescence (IF) technique. Dengue type 1-infected Vero cells were applied after fixation with cold (−20 °C) acetone.

As several hundred of Dengue cases are submitted every year, we started a programme to detect anti-Flavivirus antibodies optimally. When different antibody titers in various individuals with acute Dengue fever were compared, it turned out that high IgM antibody titers can be detected about one week after onset of infection (Fig. 3). In contrast to the specific IgM antibodies, IgG antibodies detected by IF were found only several days later. In about half of our patients the IgG antibodies were absent in the first serum specimen obtained from patients with acute Dengue fever. The HI method was more effective in detecting an early antibody response in Dengue patients, and this can be explained by our observation that the HI test detects both IgM and IgG antibodies with high sensitivity.

From our data we conclude that using only indirect immunofluorescence to diagnose acute Dengue fever, false negative results will be frequently obtained if only a single serum sample is available. Therefore, unless a second serum sample can be provided, the IF test should be applied in combination with one of the alternative methods (IgM test or HI test) which are able to detect antibodies earlier in the course of the disease. Despite that, a reliable diagnosis can be made by IF, if a significant rise in titer is demonstrable in a second serum sample.

Unfortunately all antibody tests to Flaviviruses show strong cross reactivity. Thus, it may be difficult to differentiate between anti-JB and anti-Dengue antibodies, if only serum samples and no cerebrospinal fluid is available.

In order to identify the Flavivirus type or subtype in a serum sample, we have also used RT-PCR to detect the viral RNA directly in serum samples. For our

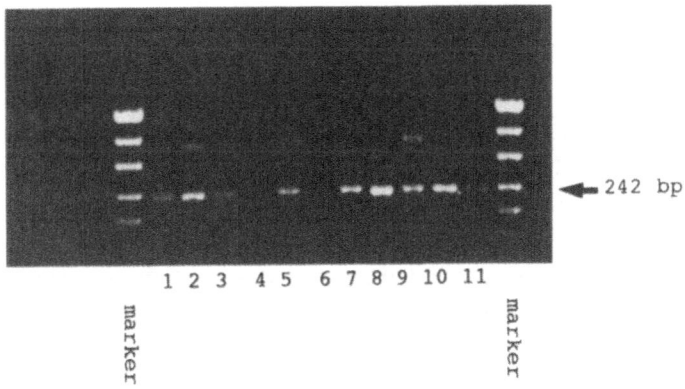

**Fig. 4.** RT-PCR using different genomic Flavivirus RNAs obtained from tissue culture supernatant (TCS) or from mouse brain (MB). *1, 2* DEN3, TCS; *3* West Nile, TCS; *4* Dengue 3, TCS; *5* Dengue 2, TCS; *6* negative control, TCS; *7* West-Nile, MB; *8* Dengue 1, TCS; *10* Dengue 4, MB; *11* Dengue 4, TCS

PCR, optimized primers have been designed showing a maximum of three mismatches with all four Dengue virus RNAs and also with West Nile virus RNA as illustrated in Fig. 4. Using these primers, the genomic RNA of all Dengue type viruses (1–4) as well as that of West Nile virus could be amplified. By PCR a DNA fragment of 235 bp was obtained with all Flaviviruses tested. In two early sera with already high IgM antibody titers viral RNA could be obtained. In both patients, a sequence compatible with that of Dengue 1 virus was obtained. Moreover, specific RNA was amplified both from cerebrospinal fluid and from serum of a patient with Japanese B encephalitis. Work is in progress to differentiate the amplified material on SSCP gels to avoid sequencing of each PCR product.

## Discussion

During our work on the diagnosis of tropical viruses we learned that numerous viral antigens have to be maintained in our laboratory for an exact diagnosis of the various clinical cases. We can now diagnose not only the common Flavivirus infections but also Lassa, Marburg or Ebola virus infections which are rarely found among travellers returning from tropical areas. In contrast, Lassa virus infections are a frequent cause of severe hemorrhagic fever in West Africa, and during our work in this region we did some epidemiological evaluations using antibody detection and a questionnaire to study the transmission of Lassa virus. From these data, we conclude that contact to the rodents and their excrements inside the houses and stored food supplies may play a role in transmission but hunting and eating of the animals seems to be even more important.

In most patients, Dengue viruses could be identified as causative agents of the flu-like tropical virus diseases, but occasionally similar symptoms were caused by Chikungunya, West Nile or Ross River viruses. Routine serological tests are not commercially available for most of these viruses; they have to be prepared in a high security lab. Unfortunately, routine tests with the exception of the relatively complicated neutralization test for Dengue virus, do not allow a differentiation between antibodies to the various Flaviviruses. Thus, using HI or immunofluorescence tests as well as our Flavi-IgM test, an acute Flavivirus infection can be diagnosed, which may be sufficient for clinicians who rarely need a type specific diagnosis. Our data show that, using indirect immunofluorescence, antibodies to Dengue virus are rarely detected during the first week after onset of symptoms. Therefore, an additional method such as the HI test or the IgM test should complement the immunofluorescence technique. For epidemiological studies, the worldwide distribution of Dengue type viruses is very important. For this purpose, the detection of the viral RNA in serum samples of the patients might be of special value [1]. The RNA sequences amplified by RT-PCR can be further characterized by sequencing or gel chromatography. Thus a type or even subtype specific identification of the Flavivirus in question would be possible.

In conclusion, our data show that due to the increased travel activities of the people in various parts of the world, an increasing number of exotic diseases must be anticipated in the future. Studies on improved diagnostic methods, on the pathogenesis and on suitable vaccines for various tropical virus diseases have a high priority in virus reseach.

## References

1. Chang GJ, Trent DW, Vorndam AV, Vergne E, Kinney RM, Mitchell CJ (1994) An integrated target sequence and signal amplification assay, reverse transcriptase-PCR-enzyme-linked immunosorbent assay, to detect and characterize flaviviruses. J Clin Microbiol 32: 477–483
2. Clos J, Brandau S (1994) pJC20 and pJC40 – two high-copy-number vectors for T7 RNA polymerase-dependent expression of recombinant genes in Escherichia coli. Prot Express Purif 5: 133–137
3. Henderson RH, Sundaresan T (1982) Cluster sampling to assess immunization coverage: a review of experience with a simplified sampling method. Bull World Health Organ 60: 253–260
4. Hufert FT, Lüdke W, Schmitz H (1989) Epitope mapping of Lassa virus nucleoprotein using monoclonal anti-nucleocapsid antibodies. Arch Virol 106: 201–212
5. Mc Cormick JB (1987) Epidemiology and control of Lassa fever. Curr Top Microbiol Immunol 134: 60–78
6. Schmitz H, Emmerich P (1984) Detection of specific IgM antibody to different flaviviruses by use of enzyme-labelled antigens. J Clin Microbiol 19: 664–667

Authors' address: Dr. H. Schmitz, Department of Virology, Bernhard-Nocht Institute for Tropical Medicine, Bernhard-Nocht-Strasse 74, D-20359 Hamburg, Federal Republic of Germany.

# Filovirus infections

Arch Virol (1996) [Suppl] 11: 77–100

# Emerging and reemerging of filoviruses

**H. Feldmann, W. Slenczka,** and **H.-D. Klenk**

Institute of Virology, Philipps-University, Marburg, Federal Republic of Germany

**Summary.** Filoviruses are causative agents of a hemorrhagic fever in man with mortalities ranging from 22 to 88%. They are enveloped, nonsegmented negative-stranded RNA viruses and are separated into two types, Marburg and Ebola, which can be serologically, biochemically and genetically distinguished. In general, there is little genetic variability among viruses belonging to the Marburg type. The Ebola type, however, is subdivided into at least three distinct subtypes. Marburg virus was first isolated during an outbreak in Europe in 1967. Ebola virus emerged in 1976 as the causative agent of two simultaneous outbreaks in southern Sudan and northern Zaire. The reemergence of Ebola, subtype Zaire, in Kikwit 1995 caused a worldwide sensation, since it struck after a sensibilization on the danger of Ebola virus disease. Person-to-person transmission by intimate contact is the main route of infection, but transmission by droplets and small aerosols among infected individuals is discussed. The natural reservoir for filoviruses remains a mystery. Filoviruses are prime examples for emerging pathogens. Factors that may be involved in emergence are international commerce and travel, limited experience in diagnosis and case management, import of nonhuman primates, and the potential of filoviruses for rapid evolution.

## Epidemiology

### Marburg hemorrhagic fever

Marburg hemorrhagic fever, Marburg, Frankfurt, Belgrade 1967

Hemorrhagic fever caused by filoviruses emerged first in 1967 (Table 1; Fig. 1). The epidemic started in mid-August with three laboratory workers of a factory in Marburg, Federal Republic of Germany, who became ill with a hemorrhagic disease after processing organs from African green monkeys (*Cercopithecus aethiops*). In the course of the epidemic 17 more patients were admitted to the hospital and two medical staff members became infected while attending the patients. The last patient who apparently had been infected by her husband during the convalescent period was admitted in November of 1967 [54, 55]. Six more cases, including two secondary infections, occurred in Frankfurt, Federal Republic of Germany, that apparently got the disease at the same time [75, 76].

**Table 1.** Outbreaks/episodes of filoviral hemorrhagic fevers

| Location | Year | Virus/subtype[a] | Cases (mortality) | Epidemiology |
|---|---|---|---|---|
| Germany/'Yugoslavia' | 1967 | Marburg | 32 (23%)[b] | Imported monkeys from Uganda source of most human infections |
| Zimbabwe | 1975 | Marburg | 3 (33%) | Unknown origin; index case infected in Zimbabwe; secondary cases were infected in South Africa |
| Southern Sudan | 1976 | Ebola/Sudan | 284 (53%) | Unknown origin; spread mainly by close contact; nosocomial transmission and infection of medical staff |
| Northern Zaire | 1976 | Ebola/Zaire | 318 (88%) | Unknown origin; spread by close contact and by use of con-taminated needles and syringes in hospitals |
| Tandala, Zaire | 1977 | Ebola/Zaire | 1 (100%) | Unknown origin; single case in missionary hospital; other cases may have occurred nearby |
| Southern Sudan | 1979 | Ebola/Sudan | 34 (65%) | Unknown origin; recurrent out-break at the same site as the 1976 outbreak |
| Kenya | 1980 | Marburg | 2 (50%) | Unknown origin; index case infected in western Kenya died, but physician secondarily infected survived |
| Kenya | 1987 | Marburg | 1 (100%) | Unknown origin; expatriate traveling in western Kenya |
| USA | 1989/90 | Ebola/Reston | 4 (0%) | Introduction of virus with imported monkeys from the Philippines; four humans asymptomatically infected |
| Italy | 1992 | Ebola/Reston | 0 (0%) | Introduction of virus with imported monkeys from the Philippines; no human infections associated |
| Ivory Coast | 1994 | Ebola/(Ivory Coast?) | 1 (0%) | Contact with chimpanzees; single case |
| Kikwit, Zaire | 1995 | Ebola/Zaire | 315 (77%) | Unknown origin; course of Outbreak as in 1976 |
| Gabon | 1995/96 | Ebola(?) | ? | Outbreak ongoing |

Beside the well documented episodes listed here, two more suspected fatal and nonfatal cases of Ebola hemorrhagic fever (see 'Ebola hemorrhagic fever 1976–1979') and a single case of Marburg hemorrhagic fever (laboratory infection) (Ryabchikova, pers. commun.) have been reported

[a]Subtypes of Marburg are not classified

[b]Numbers include a primary case which has been diagnosed some years after the epidemic (Slenczka, unpubl. data)

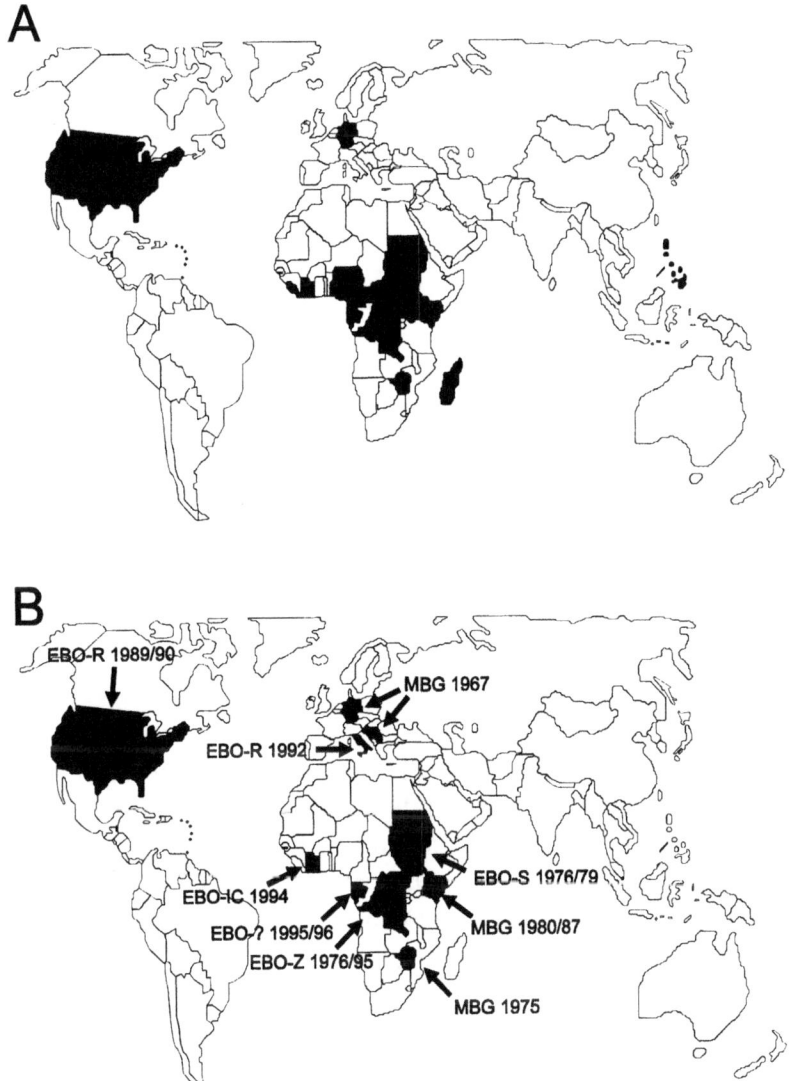

**Fig. 1. A** Prevalence of filovirus reactive antibodies. Countries are marked which have been subject to published serosurveys. For references see text ('serosurveys') and reference list. **B** Outbreaks of hemorrhagic fever caused by filoviruses. All important documented episodes are shown with the year of emergence and/or reemergence. *EBO-IC* Ebola subtype (?) Ivory Coast; *EBO-R* Ebola subtype Reston; *EBO-S* Ebola subtype Sudan; *EBO-Z* Ebola subtype Zaire; *EBO-?* Ebola subtype unknown; *MBG* Marburg

Further cases occurred in September in Belgrade, former Yugoslavia, with a veterinarian being infected performing an autopsy of dead monkeys, and his wife who nursed him during the first days of the illness [77]. Altogether, there were 31 cases, including six secondary cases, and there were seven deaths (Tables 1 and 2) [36]. Serologic data obtained some years after the epidemic suggest an additional primary case in Marburg during the 1967 outbreak (Slenczka, unpubl.

**Table 2.** Distribution of cases by transmission route during four major epidemics

| Transmission type | Germany/ 'Yugoslavia' 1967[a] | Sudan, 1976[c] | Zaire, 1976[d] | Sudan, 1979[e] |
|---|---|---|---|---|
| Nosocomial | 5 (15.6%) | 6 (2.1%) | 85 (26.7%) | 4 (11.8%) |
| Person-to-person | 1 (3.1%) | 231 (81.3%) | 149 (46.9%) | 27 (79.4%) |
| Nosocomial or person-to-person | _[f] | 18 (6.4%) | 43 (13.5%) | _[f] |
| Contact to infected monkeys | 26 (81.3%)[b] | _[f] | _[f] | _[f] |
| Neonatal | _[f] | _[f] | 11 (3.5%) | _[f] |
| Unknown | _[f] | 29 (10.2%) | 30 (9.4%) | 3 (8.8%) |
| Total number of cases | 32 (100%) | 284 (100%) | 318 (100%) | 34 (100%) |
| Total number of deaths (mortality) | 7 (21.8%) | 151 (53.2%) | 280 (88.1%) | 22 (64.7%) |

[a] Based on [56]

[b] Including one primary case which has been diagnosed some years after the epidemic (Slenczka, unpubl. data)

[c] Based on [85]

[d] Based on [86]

[e] Based on [2]

[f] No data available

data). A virus that was morphologically unique and antigenically unrelated to any known human pathogen was isolated from blood and tissues of patients by inoculation of guinea pigs and cell cultures [48, 70, 72]. The virus was named Marburg virus after the city in Germany where it was characterized first.

The infectious agent was introduced by infected monkeys imported from Uganda among which a few originally infected animals were probably responsible for the whole episode. Numbers on hemorrhagic disease and death among the monkeys from the single shipment from Uganda have never been published, but all African green monkeys experimentally inoculated with the virus died [33]. The origin of the infectious agent could be traced back to foci outside of continental Europe. During July and August vervet monkeys compounded in Entebbe (central holding station at Lake Victoria) (Fig. 2) were exported from Uganda to Germany and to former Yugoslavia via London where they potentially had contact with a large variety of animals from many parts of the world while being held in animal quarters near the airport. There had been no evidence of infection until the monkeys reached their final destination.

Aerosol transmission during the epidemic is very unlikely, and infection from monkey to man occurred by direct contact with blood or organs of the animals including the handling of tissue cultures derived thereof (Table 2). The occurrence of new cases was stopped by applying common barrier nursing techniques. Complement fixing antibodies were found in sera from some monkeys originally trapped near Lake Kyoga, the main area where vervet monkeys had been captured since the establishment of the trade in 1962. The finding of antibodies in 3 monkey trappers indicates that human infection may have occurred in Uganda during that time. However, all titers observed were weak, and an agent has never been isolated from the blood of a wild-trapped monkey or a monkey trapper [35, 36].

## Marburg hemorrhagic fever, Africa 1975–1987

Marburg virus remained an obscure medical curiosity until 1975, when three cases of Marburg hemorrhagic fever were reported from Johannesburg, South Africa (Table 1; Fig. 1) [29]. The index patient was a man who had traveled in Zimbabwe shortly before becoming ill. Seven days post-onset of his illness, his traveling companion became ill followed by a nurse who came down with the symptoms 7 days after contact with the second patient. The index case patient died 12 days post-onset of the disease, whereas both patients secondarily infected survived. An investigation was conducted along the travel route of the index patient, but no source of the virus was discovered [16, 78]. The last two episodes occurred in 1980 and 1987 in Kenya (Table 1; Figs. 1 and 2). The index patient in 1980 became ill in western Kenya and died in Nairobi. An attending physician became infected but survived. Further spread was prevented, presumably by use of barrier nursing procedures [74]. In 1987, a single fatal Marburg case was reported in western Kenya, near the location where the index patient of the 1980 episode had become infected (Johnson, unpubl. data). Both index case patients traveled in Kenya including the Mt. Elgon region (Fig. 2). This region is not far from the shores of Lake Victoria and thus close to the trapping place (Lake Kyoga, Uganda) and holding station (Entebbe, Uganda) of the monkeys that initiated the 1967 outbreak in Europe. One of the index cases had visited a cave (Kitum cave) in that area shortly before becoming ill. Serological studies in this area, however, again failed to uncover the source of the virus. These studies included an extensive investigation of many animal species inhabiting the cave (Johnson, unpubl. data).

## *Ebola hemorrhagic fever*

Hemorrhagic fever caused by Ebola virus, another filovirus, emerged in 1976, when two epidemics simultaneously occurred in Zaire and Sudan (Table 1; Figs. 1 and 2). The agent was isolated from patients in both countries and named after a small river in northwestern Zaire. This virus was morphologically similar to but serologically distinct from Marburg virus [46, 58–60, 85, 86].

## Ebola hemorrhagic fever, Sudan 1976

In June and July first cases were reported from Nzara in Western Equatoria Province of southern Sudan, a small town bordering the African rain forest zone (Fig. 2). The outbreak was strongly associated with index cases in a single cotton factory in town, and spread was to close relatives (67 cases). The epidemic was augmented by exportation of cases to neighboring areas, Maridi, Tembura, and Juba. High levels of transmission occurred in the hospital of Maridi, a teaching center for student nurses (213 cases). Despite the similarities in clinical disease

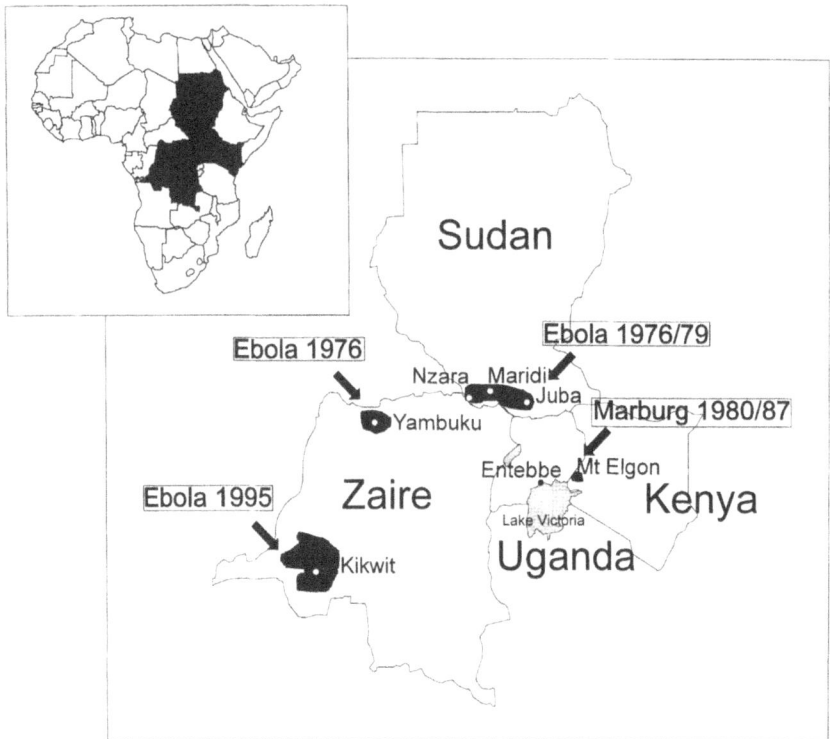

**Fig. 2.** Geographic location of the main centers of hemorrhagic fevers caused by filoviruses. Shown are the countries Kenya, Sudan, Uganda, and Zaire. The centers of the three major outbreaks of hemorrhagic fever caused by Ebola viruses are marked by black zones (Nzara, Maridi, Juba 1976; Yambuku, 1976; Kikwit 1995). Another black zone (Mt Elgon) indicates the region where two index cases of Marburg hemorrhagic fever became infected (1980/87). A central station for wild-caught monkeys in the late 1960's was located near Entebbe at Lake Victoria. Infected vervet monkeys were shipped from here to Germany and the former Yugoslavia in 1967

and mortality rates, the epidemic in Nzara contrasted with Maridi. The Nzara outbreak mostly involved factory workers and their close relatives, whereas in Maridi the hospital served both as focus and amplifier of the infection. At least a third of the staff of Maridi hospital was infected of which 41 died. The outbreak lasted until November and comprised approximately 15 generations of person-to-person transmissions (Table 2). Transmission of the disease required close contact with an acute case and was usually associated with nursing patients. The overall secondary attack rate was 12% and documented the relatively slow rate of spread into the community outside the hospital. In total, there were 284 probable and confirmed cases involved of which 151 died (53%) (Tables 1 and 2). Establishment of strict barrier nursing and classic public health principles, and identification and isolation of cases were successful in controlling the epidemic. The episode in Nzara died out spontaneously [1, 7, 28, 73, 85].

## Ebola hemorrhagic fever, Zaire 1976

By the end of August a second epidemic started in equatorial rain forest areas of northern Zaire (Fig. 2). A direct link between the two epidemics has always been discussed but could never be verified. In total, there were 318 probable or confirmed cases and 280 deaths (88%) (Tables 1 and 2). The presumable index case came to Yambuku Mission hospital for treatment of acute malaria, where he received an injection of chloroquine. It remains unclear whether this man was the source of the epidemic or became infected by the injection. Most persons acquired the disease following contact with patients, but for more than 25% the only apparent risk factor consisted in receipt of injections given at Yambuku Mission hospital (Table 2). Nearly all survivors were infected by person-to-person contact. All ages and both sexes were affected, but the highest incidence was in women aged 15 to 29 years, who were frequently patients attending antenatal and outpatient clinics at the hospital. Of the 17 medical staff members of Yambuku Mission hospital, 13 acquired the disease and 11 died. Although transmission occurred predominantly in the outpatient clinics of the hospital, there was subsequent dissemination of disease in surrounding villages to people caring for sick relatives, attending child birth, or other forms of close contact. The overall secondary attack rate was approximately 5% but amounted to about 20% in close relatives of a patient. The epidemic, which lasted from the end of August until the end of October, spread relatively slowly in the area, and all infected villages (55; population < 5 000) were located within 60 km of Yambuku. The epidemic ended after institution of basic quarantine procedures. The major mode of spread, contaminated syringes and needles, almost completely terminated when the hospital closed [7, 8, 46, 86].

## Ebola hemorrhagic fever 1976–1979

One Ebola hemorrhagic fever case occurred in the United Kingdom in 1976, in a laboratory worker who pricked his finger with a needle while transferring homogenized liver of a guinea pig infected with the new African Ebola virus. The case patient survived, perhaps because of treatment with human leukocyte interferon and human convalescent plasma [20]. In 1977, another confirmed, fatal case of Ebola hemorrhagic fever was reported from Tandala, Zaire, about 325 km from the original focus of the 1976 outbreak (Table 1) [37]. Reports on two other probable fatal cases and a serologically confirmed nonfatal case, which had occurred in 1972 in a missionary physician after he had performed an autopsy on a patient diagnosed clinically as having yellow fever [47, 59, 79], are controversially discussed.

In 1979, Ebola hemorrhagic fever reemerged in Nzara and Yambio which are located in the remote savanna of southern Sudan, near the border with Zaire (Table 1; Figs. 1 and 2). The index case, a 45-year old man, was admitted to the Nzara hospital with fever, vomiting and diarrhea. He developed gastrointestinal bleeding and died three days post admission. The index case worked in the same textile factory cited as the source of the 1976 outbreak in Sudan. The outbreak took place from July 31 to October 6, 1979, and started in the hospital where the index case patient was responsible for four nosocomial infections, which in turn led to disease in five families. Thirty-three cases could be traced to a human source of infection with 22 fatalities (65% mortality) (Tables 1 and 2). Seven generations of virus transmission were estimated and mortality changed from 89% in the first 4 generations to 38% in the last three. Studies within families confirmed reports from previous outbreaks (Sudan and Zaire, 1976), and suggested that Ebola virus is not easily transmitted. Again, the hospital appeared to be the important focal point for dissemination of the disease. Follow-up studies failed to identify the ecology of the virus [2, 87].

## Reston hemorrhagic fever

In 1989 veterinary staff in a primary import quarantine facility in Reston, Virginia, noted numerous deaths among cynomolgus monkeys in one animal room and suspected simian hemorrhagic fever (SHF) as the cause (Table 1; Fig. 1). Samples tested in the virus laboratory yielded SHF, but also were shown to contain a filovirus. A virus closely related to Ebola virus was isolated from monkeys, it was called Ebola Reston [13, 18, 42]. The cynomolgus monkeys (*Macaca fascicularis*) had been imported from the Philippines. The shipment arrived either via Amsterdam or directly from the Philippines across the Pacific ocean. No link to African or animals of other continents could be established on any route. Therefore, the presumption prevails that this new Ebola virus isolate is of Asian origin. The role of SHF in initiating or propagating the epidemic is unknown, but the new filovirus was found to be pathogenic for monkeys under experimental conditions. Filoviral antigen and particles were found in tissues of naturally and experimentally infected monkeys in close anatomic relationship to the pathologic lesions [18, 26, 30, 42]. The epizootic occurring in monkeys spread through affected rooms by droplet contact between animals in adjacent cages or to distant cages and different rooms by larger droplets and/or small particle aerosols [62, 63]. The airborne route of transmission was supported by the prominent respiratory involvement of the infected monkeys. Spread of the disease to other rooms within the facility led to a decision to euthanize all the monkeys in the building. Resumption of importation of monkeys led to new outbreaks of disease. Subsequent investigation traced the source of the infection to a single holding facility in the Philippines that was thought to have furnished all identified infected shipments, including monkeys sent to facilities in Texas and Pennsylvania [34]. Four animal handlers at the quarantine facility became infected as judged by serological tests and, in one case, by virus isolation. All four had a high level of daily exposure, but except for one, who cut himself while performing a necropsy, the mode of transmission remained unclear [15]. None of them had an unexplained febrile illness suggesting that this virus may be less pathogenic for humans in contrast to human infections with previously known filoviruses resulting in significant disease and mortality rates ranging from

25–90%. However, the observations should not be interpreted as assuring that this new virus is avirulent for humans.

In 1992, cynomolgus monkeys were imported into Italy from the same holding compound in the Philippines which exported the monkeys causing the 1989–1990 epizootic (Table 1; Fig. 1). Ebola Reston virus was isolated from three monkeys which died. No illness in associated humans was reported [89]. Reportedly, in 1991 monkeys were eliminated from the holding facility and cages were disinfected. Thus, the virus had either persisted or was re-introduced by a similar mechanism as lead to the 1989–1990 epizootic.

Evidence for ongoing epizootic and transmission of Ebola viruses among captured monkeys at the export facility in the Philippines was obtained in 1990 and 1993 (Fig. 1). In 1990, filoviral antigen was detected by ELISA in 52.8% of monkeys dying within the facility, but never in dead monkeys from another facility in the Philippines. The investigation suggested that the type of holding cage was important in transmission, since being in gang cage at the time of the initial serosurvey was a significant risk factor for subsequent infection [34]. Again in summer 1993, high titered ELISA antibodies were present in monkeys held at that facility, but no evidence of viral antigen was found. Monkeys imported at that time from the facility into the United States had stable IgG titers suggesting infection in the recent past but not during quarantine [64]. Even so, the original source remains unknown, it seems likely that naturally infected wild monkeys captured in the Philippines are the source of the virus.

*Emerging and reemerging of Ebola hemorrhagic fever in Africa*

Ebola hemorrhagic fever, Ivory Coast 1994

Two episodes of mortality were noticed among a troop of chimpanzees in 1992 (8 deaths) and 1994 (12 deaths). The chimpanzees were objects of a 15 year observation by ethnologists in the Tai National Park in western Cote-d'Ivoire. Several of the dead animals showed signs of hemorrhages, and one of the animals was autopsied in the field. A 34-year old woman developed a dengue-like syndrome 8 days after performing the autopsy. She was admitted to the hospital in Abidjan two days later with continuing fever resistant to anti-malaria treatment, diarrhea, and pruritic rush. The evacuation to Switzerland followed 5 days later when she developed a syndrome similar to that described for surviving Ebola-infected patients; she recovered without sequelae. An infection with an Ebola virus was confirmed by isolate-specific IgM and IgG antibodies, Ebola-Zaire-specific IgG antibodies, antigen ELISA, reactivity to an Ebola serotype-specific monoclonal antibody, and virus isolation. The isolated agent was antigenically and genetically distinct from all previously known Ebola isolates and most closely related to the Zairian subtype [53, 69]. Epidemiological data suggest an Ebola epizootic among the troop of chimpanzees as the cause of death. Contact with infectious blood and tissues during necropsy was considered to be the most likely source of the human infection. Organs of the dead chimpanzee were studied by immunohistochemistry and the findings were similar to those seen in material of the 1976 Ebola outbreaks and monkeys experimentally infected with Ebola virus. None of the persons in contact with either the case patient or the materials of the chimpanzee developed the disease or were tested antibody-positive (Table 1; Fig. 1) [53].

## Ebola hemorrhagic fever, Zaire 1995

Recently Ebola hemorrhagic fever reemerged in Zaire. The first identified case related to the outbreak had onset of illness on January 6, 1995 (Table 1; Figs. 1 and 2). Until August 24, the official end of the epidemic, 315 cases occurred and 244 patients died (77%). The center of the epidemic was Kikwit, a city with a population of approximately 400 000 and two hospitals (Kikwit General hospital and Kikwit 2), and the surrounding areas in Bandundu region in southwestern Zaire. The first case at Kikwit General hospital was a male laboratory worker who had previously been admitted to Kikwit 2 hospital. A laparatomy was performed after a differential diagnosis of typhoid fever with intestinal perforation. This was followed by a second laparatomy which showed massive intra-abdominal hemorrhage. The patient died 3 days later. Four days after the first laparatomy the first case among medical staff members occurred with fever, headache, muscle aches, and hemorrhages. About three-quarters of the first 70 patients within the subsequently developing epidemic were health care workers. Prior to this time, cases have been sporadic. Major risk factors for contracting disease were involvement in patient care in hospitals, and households and preparation of bodies for burials. This is reflected by the fact that 26% of the cases with known professional occupation were medical staff members or students and 21% were housewives. During the course of surveillance, several chains of infections could be traced back as far as late December 1994. The chain of the presumable index case, a charcoal worker, involves 7 out of 12 persons living in his household. In order to define the natural reservoir of the virus, field teams captured birds, mammals, and thousands of possible insect vectors mainly within the area around the working place of the presumable index case, but also at different places in the close vicinity of Kikwit. Up to now no evidence of the reservoir could be found, but testing continues ([90, 91], CDC, pers. comm.). The virus isolated from case patients was antigenically and genetically closely related to the 1976 Zairian isolate of Ebola [69].

## Transmission

The usual pattern seen in large outbreaks of filovirus hemorrhagic fever begins with a focus that disseminates infection to several patients. Secondary and subsequent infections occur in close family members or among medical staff. The epidemic terminates, because the index focus is transient and the spread of the virus is inefficient (Table 2).

Person-to-person transmission by intimate contact is the main route of infection in human filoviral hemorrhagic fever outbreaks. However, more superficial contact and sleeping in the same room has a relatively low attack rate, whereas nursing of patients and preparing bodies for burials increase the chances of becoming infected. In the Ebola outbreak of 1976 (27%), and to some extent 1995, nosocomial transmission via contaminated syringes and needles has been a major problem. Neonatal transmission has been reported from the 1976 outbreak in Zaire (Table 2). Additional late spread via semen has been documented in Marburg disease [54]. In general, transmission seems to be not efficient as documented by secondary attack rates which in average rarely exceed 10–12%. Among family members, however, secondary attack rates can increase as reported during the Ebola outbreaks in 1976.

Based on the experience of former and recent episodes isolation of patients and use of strict barrier nursing procedures (e.g., gowns, gloves and masks) are required and sufficient to interrupt transmission. Transmission by droplets and small aerosols has been observed with the Ebola Reston epizootic [63]. Cage-to-cage transmission of monkeys experimentally infected with Marburg or Ebola viruses also suggest aerosol transmission ([65]; Johnson, unpubl. data), and monkeys exposed to aerosols of filoviruses get readily infected ([4]; Johnson, unpubl. data). Furthermore, virions were detected in alveoli of infected animals [63] and humans [93]. Course of human outbreaks, however, indicate that aerosols and droplets do not seem to be an important route of transmission, although there are occasionally secondary cases without any clear history of close contact to infected individuals [64]. Information regarding case management and approaches to minimize virus spread in critical situations such as outbreaks have been published [11, 27, 63, 88, 92].

Wild monkeys are an important source for the introduction of filoviruses which has clearly been demonstrated in 1967 for Marburg [56], in 1989–1990 and 1992 for Ebola Reston [42, 89], and in 1994 for Ebola Ivory Coast [53]. Quarantine of imported non-human primates and professional handling of animals are essential to prevent introduction of these agents into human (for guidelines see [14]).

*Natural reservoir*

Ecological investigations have been conducted after almost all filovirus outbreaks. It has generally been possible to identify an index human case or index group of imported non-human primates, but the origin in nature and the natural history of Marburg and Ebola viruses remain a mystery. It is generally believed that the viruses are zoonotically transmitted to humans from ongoing life cycles in animals or arthropods. Species such as guinea pigs, primates, bats, and hard ticks have been discussed as natural hosts, but a high frequency of false-positive results, especially when the Ebola indirect immunofluorescence assay (IFA) was used to detect filovirus-specific antibodies, has contributed to difficulties in interpretation. Thus, all attempts to backtrack from human index cases in Africa or from epidemics in monkeys in Africa and the Philippines failed to uncover a reservoir.

The Ebola Reston epizootic again raised the question of whether non-human primates might be the primary reservoir. This seems to be not very likely, since all filoviruses are relatively pathogenic for experimentally and naturally infected animals and persistence of virus in such animals has never been demonstrated [26]. The emergence of the presumably new Ebola subtype 'Ivory Coast' demonstrated for the first time a connection between a human infection and naturally infected monkeys in Africa [53]. But again, based on the ongoing deaths in the chimpanzee colony, it seems to be unlikely that these monkeys are the natural reservoir of this virus. Bats have seriously been discussed as sources of infection for two index Marburg cases (1980 and 1987) and for an index case in one of the Sudan outbreaks (Table 1). Despite failure to detect filovirus in the respective environments, the possibility still exists that an aerosol transmission cycle in bats or other mammals plays some role in the natural history of filoviruses.

All attempts to identify the reservoir of filoviruses were handicapped by the lack of sensitive diagnostic tools. With the enzyme immunoassays, antigen detection assays, and PCR assays developed during the past years ([24, 52, 71]; Ksiazek, unpubl. data) the situation has now improved, presupposed appropriate material can be collected. A major investigation has recently been started in association with the latest reemergence of Ebola subtype Zaire in Kikwit.

## Serological studies

Serological studies have been performed over the years in various geographical regions of African countries (Fig. 1) [2, 3, 6, 8, 28, 32, 35–37, 40, 41, 43–45, 47, 56, 57, 59, 71, 73, 74, 78, 80–82]. Results are mainly based on the indirect immunofluorescence technique which, however, is prone to yield false-positives, but a technique that can be easily applied under field conditions. Thus, many of the data are of limited reliability, but helpful to demonstrate that endemic areas are mainly located in the central African region (Fig. 1). Recent serosurveys using additional techniques (immunoblot, enzyme linked immunosorbent assay) and conducted in countries outside of Africa, such as Germany [5], the United States [15], and the Philippines [34] indicate filoviral activity in those countries as well, and may suggest that presently known or unknown filoviruses are also endemic outside of Africa. The value of those serosurveys is controversially discussed, and none of the serological studies was verified by virus isolation in association with the surveys. The fact that filoviruses belong to the order *Mononegavirales*, an order combining many common human pathogens that could be the cause of crossreactivity, has to be considered in this discussion.

## The infectious agents

### Virion morphology and structure

In most virus preparations, Marburg and Ebola virions are pleomorphic, appearing as either long filamentous and sometimes branched forms, or in shorter "U"-shaped, "6"-shaped, or circular configurations (Fig. 3). The filamentous forms vary greatly in length (up to 14 000 nm), but the unit length associated with peak infectivity is 790 nm for Marburg virus and 970 nm for Ebola virus. Virions have a uniform diameter of 80 nm. They are composed of a helical nucleocapsid (50 nm in diameter, with an axial space 20 nm in diameter and a helical periodicity of about 5 nm). The nucleocapsid consists of a noninfectious, negative-sense, linear single-stranded RNA molecule and of four of the virion structural proteins: nucleoprotein (NP), virion structural proteins (VP) 35 and 30, and the polymerase (L) protein. The nucleocapsid is surrounded by a lipid envelope derived from the host cell plasma membrane and a surface projection layer composed of trimers of the viral glycoprotein (GP). Two additional viral proteins, VP40 and VP24, are membrane-associated (Table 3; Fig. 3) [19, 22, 25, 31, 49–51, 58, 61, 66, 70, 83].

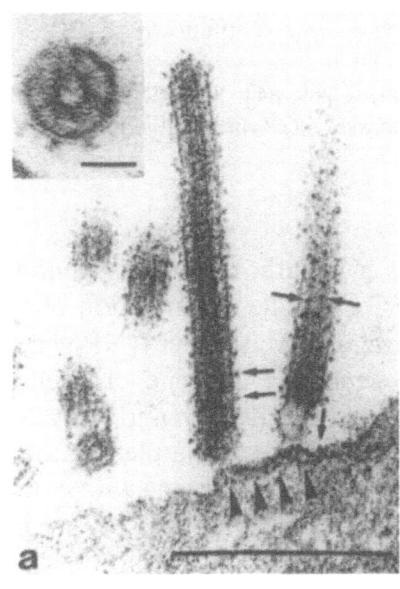

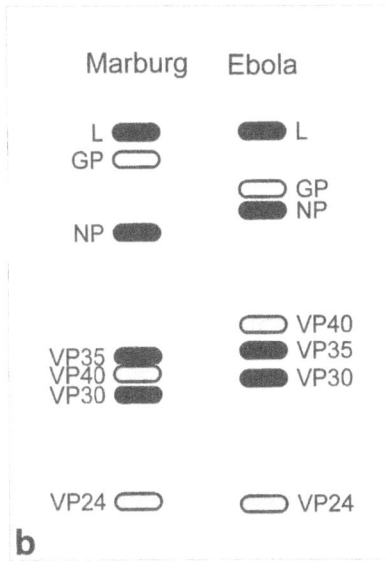

**Fig. 3.** Morphology of filoviral particles. **a** Electron micrograph. Budding of Marburg virus particles from the plasma membrane of infected primary cultures of human endothelial cells. Particles consist of a nucleocapsid surrounded by a membrane in which spikes are inserted (arrows). The nucleocapsid contains a central channel (inset). The plasma membrane of infected cells is often thickened at locations where budding occurs (arrowheads). Ultrathin section – bar: 0.5 μm; bar inset: 50 nm. **b** Electrophoretic mobility patterns of filoviral structural proteins. The mobility patterns (SDS-PAGE) of structural proteins of Marburg and Ebola type viruses are compared. Four proteins are involved in the nucleocapsid formation; polymerase or large (*L*) protein, nucleoprotein (*NP*), virion structural protein (*VP*) 30 and VP35 (black). The glycoprotein (*GP*) is a transmembrane protein and anchored with the carboxy-terminal part in the virion membrane. Homotrimers of GP form the spikes on the virion surface (arrows in **a**). VP40 and VP24 are membrane-associated proteins

**Table 3.** Filoviral proteins and their proposed function

| Designation | Virus type | Encoding gene | Localization | Proposed function |
|---|---|---|---|---|
| NP | MBG/EBO | 1 | ribonucleocapsid complex | encapsidation |
| VP35 | MBG/EBO | 2 | ribonucleocapsid complex | phosphoprotein analogue |
| VP40 | MBG/EBO | 3 | membrane-association | *matrix* protein |
| GP | MBG/EBO | 4 | surface (transmembrane protein) | receptor binding, fusion |
| VP30 | MBG/EBO | 5 | ribonucleocapsid complex | encapsidation, RNA binding |
| VP24 | MBG/EBO | 6 | membrane-association | unknown |
| L | MBG/EBO | 7 | ribonucleocapsid complex | RNA-dependent RNA polymerase |
| sGP | EBO | 4[a] | nonstructural, secreted | unknown |

[a] Expressed by RNA editing and/or translational frameshifting [69, 84]; *NP* nucleoprotein; *VP* virion structural protein; *GP* glycoprotein; *L* large protein (polymerase); *sGP* small glycoprotein; *MBG* type Marburg filoviruses; *EBO* type Ebola filoviruses

The genome is approximately 19 kilobases in length, the largest genome reported for members of the order Mononegavirales. The organization of the seven genes is sequential and shows the following order: 3'leader – NP gene – VP35 gene – VP40 gene – GP gene – VP30 gene – VP24 gene – L gene – 5'trailer (Fig. 4). The extragenic sequences at the extreme 3' (leader) and 5' (trailer) ends of the genomes are conserved and show a high degree of complementarity. Genes are delineated by conserved transcriptional signals, beginning with a start site at the 3' end and terminating with a stop (polyadenylation) site. Beside common characteristics, there are others that distinguish filovirus genomes from those of rhabdoviruses and paramyxoviruses: (i) transcriptional signals of filoviruses contain a common sequence (3' UAAUU, at the 5' end of start sites and at the 3' end of stop sites), (ii) genes possess long noncoding regions at their 3' and/or 5' ends, and (iii) localization of overlapping genes [10, 21, 22, 68].

An unusual feature of the organization and transcription of the GP genes of Ebola viruses is the fact that they are encoded in two open reading frames. The full-length GP is expressed only through transcriptional editing [69, 84] or translational frameshifting [69] close to the editing side (Fig. 4). The primary gene product of the GP gene is a small nonstructural glycoprotein (sGP) of approximately 60 kDa that is secreted from infected cells in large quantities [69, 84].

### Classification

In early days, the morphology of virion particles has led to proposals to classify filoviruses in the family *Rhabdoviridae*. In 1982, the classification in a separate family, called *Filoviridae*, based on unique morphologic, biochemical and

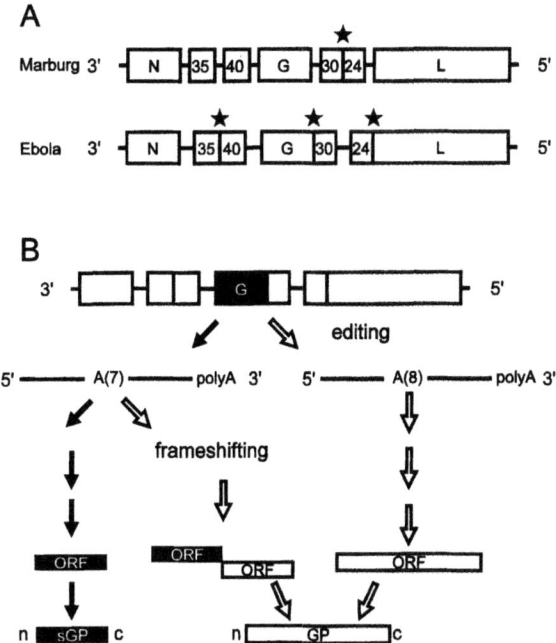

**Fig. 4. A** Genome organization of filoviruses. Filoviral genomes consist of a single, negative-stranded, linear RNA molecule. Differences in organization between Marburg and Ebola type viruses are indicated. Asterisks show the positions of gene overlaps. **B** Expression strategies of gene 4 of type Ebola viruses. Gene 4 of Ebola viruses is transcribed from two open reading frames (*ORF*s). The primary gene product is a small glycoprotein (*sGP*). Full-length glycoprotein (*GP*) can be expressed by two independent mechanisms; RNA editing and/or translational frameshifting. *A* Adenosine residue; *c* carboxy-terminal end of proteins; *n* amino-terminal end of proteins; *G* glycoprotein gene; *L* polymerase (L) gene; *N* nucleoprotein gene; 24/30/35/40, virion structural protein (VP) genes; *3', 5'* terminal ends of genomes and subgenomic RNAs

physicochemical features was introduced [50]. Today all nonsegmented negative-stranded RNA viruses are grouped in the order *Mononegavirales* bearing three distinct families, *Paramyxoviridae, Rhabdoviridae,* and *Filoviridae* [39]. All these viruses share a similar genomic organization with conserved regions at both ends encoding the core and polymerase (L) proteins and surrounding a variable part in the middle encoding the envelope proteins. Filovirus genomes are more complex than those of lyssaviruses and vesiculoviruses and align organizationally more closely to members of the genera *Paramyxovirus* and *Morbillivirus.* This is also supported by amino acid sequence comparison of different structural proteins (e.g., NP and L protein) [22]. Transcription and replication of all these viruses follow similar mechanisms and take place in the cytoplasm of the infected cell.

The family *Filoviridae* consists of a single genus, *Filovirus,* which can be separated into two types, Marburg and Ebola. Nucleotide sequence comparison among Marburg and Ebola viruses shows only scattered similarities which is in contrast to similarities seen among amino acid sequences of structural proteins.

This finding indicates that these agents may have diverged at some point in the distant past. The main type-specific differences are:

(1) *Lack of serologic cross-reactivity* [23, 51, 52, 67]. Between the two filovirus types serological cross-reactivity does not exist. Subtypes of Ebola share common epitopes, and GP-specific polyclonal antibodies react type-specific.

(2) *Protein profiles of virion structural proteins on SDS-PAGE* [19, 22, 23, 51]. Sodium dodecylsulfate-polyacrylamid gel electrophoresis (SDS-PAGE) profiles clearly distinguish Marburg from Ebola-type viruses (Fig. 3). Differences are evident for GP, which migrates much slower with Marburg viruses, and for NP and VP40, which migrate faster with those viruses. In addition, minor differences are found in the migration patterns of VP24. Within the Ebola type, subtypes can be differentiated by the mobility patterns of the NP and VP40. Subtype Zaire viruses possess a slower migrating NP and a faster migrating VP40 compared with the isolates of the Sudan and Reston subtype. Reston, on the other hand, can be distinguished from Zaire and Sudan subtype isolates by a larger VP40. In contrast, Marburg viruses are homogenous in their SDS-PAGE profiles.

(3) *Sialylation pattern of carbohydrates* [23, 25, 31]. Differences in terminal sialylation of carbohydrates are found following propagation of filoviruses in Vero cells, clone E6, and MA 104 cells. Both cell lines originate from monkey kidney and are commonly used for filovirus isolation. Marburg viruses completely lacks sialic acid when propagated in those cells.

(4) *Genetic makeup* [10, 21, 22, 68, 69, 83, 84]. The 3′ noncoding region of the NP gene is much shorter with Marburg than with Ebola viruses. Positions and numbers of gene overlaps are different and Ebola-type viruses possess more than one overlap (e.g., Zaire subtype between genes VP35/VP40, GP/VP30, and VP24/L), whereas Marburg viruses have only one overlap between genes VP30/VP24. In contrast to Ebola viruses, there is no evidence for expression of a nonstructural glycoprotein with Marburg viruses, and RNA editing has not been demonstrated with those viruses (see above). On the other hand, a second overlapping open reading frame is highly conserved in the VP35 gene of Marburg viruses, but is not present in Ebola-type viruses (Feldmann, unpubl. data). An expression product has not yet been identified.

Filoviruses are classified as "*Biological Level 4*" agents (WHO; Risk Group 4) based on their high mortality rate, person-to-person transmission, potential aerosol infectivity, and the lack of vaccines and chemotherapy. Maximum containment is required for all laboratory work with infectious material [12, 88, 92].

### Genetic variability

In general, there is little genetic variability among viruses belonging to the Marburg type of filoviruses and known subtypes do not exist. Recent sequence data on parts of the second (VP35) and fourth (GP) open reading frame, however, indicate the coexistence of at least two different genetic lineages. One lineage comprises isolates from the 1967 epidemic which are very closely related, and viruses isolated from the episodes in Zimbabwe (1975) and Kenya (1980). The second lineage is represented by the isolate from the 1987 Kenyan case which shows more than 20% divergence to the others ([51]; Feldmann, unpubl. data). Ebola, however, can be subdivided into at least three subtypes – *Zaire, Sudan, and Reston* [22, 23, 68]. A distinction within the Ebola type had previously been based on peptide and oligonucleotide mapping [9, 17]. This has recently been confirmed by sequence comparison. Molecular characterization of the 'Ivory Coast' agent now points to the existence of a fourth subtype [69]. All subtypes are approximately equally distinct from each other in their nucleotide sequence. The genetic variability within the subtypes is low and comparable with that seen for most of the Marburg type viruses. The fact that Ebola Reston does not represent a different lineage may question the Asian origin of this virus, and would support its recent introduction from Africa. The similarity of Ebola Zaire 1976 and 1995 is striking and unexpected. High similarity among filoviruses temporally and geographically isolated far apart from each other suggests that variants may emerge with comparatively low frequencies in nature and that filoviruses may occupy specific niches. The genetic variability among filoviruses in general is less than that of many other RNA viruses.

## Factors of emergence/reemergence

Filovirus are among the most pathogenic human viruses. Yet, we are only beginning to understand the interactions of these viruses with their hosts, and our knowledge on genetics, pathogenicity, and natural history is still limited. Even though outbreaks in human and non-human primates have so far always been self-limiting, our ignorance concerning the natural reservoir, the potential of these viruses to be transmitted by aerosol, and the lack of immunoprophylactic and chemotherapeutic measures make these infections a matter of high concern in biomedical science. The chronology of human epidemics and epizootic in non-human primates proves that filoviruses are prototypes of emerging and/or reemerging pathogens. The recent news of the reemerging of Ebola subtype Zaire in Kikwit alarmed the world which was already sensitized by books, movies, and television reports. Several factors have to be discussed as for the emergence/reemergence of hemorrhagic fever caused by filoviruses.

(1) *International commerce and jet travel.* A major public health concern for many countries in the world has been the potential of spread as a result of international commerce and jet travel. Infected persons can be at nearly any point in the world within the incubation period and can become the starting case of a new epidemic in even a non-endemic area.

(2) *Transmission route.* Inter-human spread in the course of normal social interactions as well as during sexual intercourse has been demonstrated. In previous human epidemics, transmission was primarily by contaminated needles and syringes or close contact with case patients, although some degree of spread to contacts with less intimate interactions was reported. An uncontrolled

propagation from person to person in an urban environment would not be likely. Fortunately, to date all outbreaks of filoviral hemorrhagic fever have been self-limiting. Aerosol transmission as an important route for spread is doubtful but cannot be excluded based on recent data obtained from experimentally and naturally infected monkeys and human cases.

(3) *Limited knowledge concerning genetics and pathogenesis of the agents.* Investigations of filoviruses using modern technologies started approximately 10 years ago. We are still just at the beginning to understand basic mechanisms in biology and pathogenesis of filoviruses.

(4) *Limited experience in diagnosis and case management.* Filoviruses are classified as biosafety level 4 agents and therefore, work with the agents and to a high extent also diagnosis can only be performed in a few selected laboratories. More sensitive and safe diagnostic tools have been recently developed and, hopefully, can be provided in the near future. Medical staff has to be trained in proper management and care of patient cases, and national or international units have to be founded for emergency situations. National public health systems have to be prepared for emergency situations and distribution of information to physicians and the population has to be provided.

(5) *Import of nonhuman primates.* Non-human primates are still a vital source to the biomedical community and needed for the development, safety testing, and production of viral vaccines, and as important models for studying human diseases. Proper quarantine procedures should help to minimize the risk of introduction by wild-caught monkeys (see [14]). In addition, breeding of monkeys in colonies instead of importation will minimize the risk of introduction of these agents.

(6) *Unknown reservoir.* The origin of filoviruses in nature is still unknown. Knowledge of the natural reservoir would be helpful to develop strategies to prevent infection.

(7) *Potential of evolution.* Filoviruses, like other RNA viruses, presumably have a potential for evolution due to an error rate of the virus-encoded polymerase and a lack of repair mechanisms [38]. The consequence may be a spectrum of genetic variants which are selected by the hosts for different transmissibility, virulence and other biological properties. Sequence analysis of different filovirus isolates, however, suggest a low frequency for the emergence of variants which is in contrast to many other RNA viruses. Changes in socio-economical structures, such as increase in human population, increase in speed, variety and frequency of travel, and disruption of social structures, may support the development of mutant virus populations and the probability of a filovirus truly emerging as a serious public health problem.

## Acknowledgements

The authors greatly acknowledge Dr. Anthony Sanchez, Dr. Clarence J. Peters (Centers for Disease Control and Prevention, Atlanta, Georgia, U.S.A.) and Dr. Viktor E. Volchkov (Institute of Virology, Philipps-University, Marburg, Federal Republic of Germany) for their helpful suggestions and discussions. The work on filoviruses in the Institute of Virology,

Philipps-University of Marburg, Germany, is presently and has previously been supported by the Deutsche Forschungsgemeinschaft (DFG), grants KL 238/1-1, SFB 286, Fe 286/4-1.

# References

1. Babiker Mohd el Tahir (1978) The haemorrhagic fever outbreak in Maridi, western Equatoria, southern Sudan. In: Pattyn SR (ed) Ebola virus hemorrhagic fever, 1st ed. Elsevier/North-Holland, Amsterdam, pp 125–127

2. Baron RC, McCormick JB, Zubeir OA (1983) Ebola hemorrhagic fever in southern Sudan: hospital dissemination and intrafamilial spread. Bull World Health Organ 61: 997–1003

3. Bauree P, Bergmann JF (1983) Ebola virus infection in man: a serological and epidemiological survey in the Cameroon. Am J Trop Med Hyg 32: 1465–1466

4. Bazhutin NB, Belanov EF, Spiridonov VA, Voitenko AV, Krivenchuk NA, Krotov SA, Omelchenko NI, Tereschenko AY, Khomichev VV (1992) The influence of the methods of experimental infection with Marburg virus on the features of the disease process in green monkeys. Vopr Virus 37: 153–156

5. Becker S, Feldmann H, Will C, Slenczka W (1992) Evidence for occurrence of filovirus antibodies in humans and imported monkeys: do subclinical filovirus infections occur worldwide? Med Microbiol Immunol 181: 43–55

6. Blackburn NK, Searle L, Taylor P (1982) Viral haemorrhagic fever antibodies in Zimbabwe schoolchildren. Trans R Soc Trop Med Hyg 76: 803–805

7. Bowen ETW, Lloyd G, Harris WJ, Platt GS, Baskerville A, Vella EE (1977) Viral haemorrhagic fever in southern Sudan and northern Zaire. Lancet 1: 571–573

8. Breman JG, Piot P, Johnson KM, White MK, Mbuyi M, Sureau P, Heymann DL, van Nieuwenhove S, McCormick JB, Ruppol JP, Kintoki V, Isaäcson M, van der Groen G, Webb PA, Ngvete K (1978) The epidemiology of Ebola haemorrhagic fever in Zaire, 1976. In: Pattyn SR (ed) Ebola virus hemorrhagic fever, 1st ed. Elsevier/North-Holland, Amsterdam, pp 103–124

9. Buchmeier MJ, DeFries R, McCormick JB, Kiley MP (1983) Comparative analysis of the structural polypeptides of Ebola virus from Sudan and Zaire. J Infect Dis 147: 276–281

10. Bukreyev AA, Volchkov VE, Blinov VM, Dryga SA, Netesov SV (1995) The nucleotide sequence of the Popp (1967) strain of Marburg virus: a comparison with the Musoke (1980) strain. Arch Virol 140: 1589–1600

11. Centers for Disease Control and Prevention (1988) Management of patients with suspected viral hemorrhagic fever. MMWR 37 [Suppl 3]: 1–16

12. Centers for Disease Control and Prevention (1993) Biosafety in microbiology and biomedical laboratories. US Department of Health and Human Services (HHS), publication No. (CDC) 93-8395, US Government Printing Office, Washington

13. Centers for Disease Control and Prevention (1989) Ebola virus infection in imported primates – Virginia, 1989. MMWR 38: 831–832, 837–838

14. Centers for Disease Control and Prevention (1990) Update: Ebola-related filovirus infection in nonhuman primates and interim guidelines for handling nonhuman primates during transit and quarantine. MMWR 39: 22–24, 29–30

15. Centers for Disease Control and Prevention (1990) Update: filovirus infection associated with contact with nonhuman primates or their tissues. MMWR 39: 404–405

16. Conrad JL, Isaäcson M, Smith EB, Wulff H, Crees M, Geldenhuys P, Johnston J (1978) Epidemiologic investigation of Marburg virus disease, Southern Africa, 1975. Am J Trop Med Hyg 27: 1210–1215

17. Cox NJ, McCormick JB, Johnson KM, Kiley MP (1983) Evidence for two subtypes of Ebola virus based on oligonucleotide mapping of RNA. J Infect Dis 147: 272–275

18. Dalgard DW, Hardy RJ, Pearson SL, Pucak GJ, Quander RV, Zack PM, Peters CJ, Jahrling JB (1992) Combined simian hemorrhagic fever and Ebola virus infection in cynomolgus monkeys. Lab Anim Sci 42: 152–157

19. Elliott LH, Kiley MP, McCormick JB (1985) Descriptive analysis of Ebola virus proteins. Virology 147: 169–176

20. Emond RTD (1978) Isolation, monitoring and treatment of a case of Ebola virus infection. In: Pattyn SR (ed) Ebola virus hemorrhagic fever, 1st ed. Elsevier/North-Holland, Amsterdam, pp 27–32

21. Feldmann H, Mühlberger E, Randolf A, Will C, Kiley MP, Sanchez A, Klenk H-D (1992) Marburg virus, a filovirus: messenger RNAs, gene order, and regulatory elements of the replication cycle. Virus Res 24: 1–19

22. Feldmann H, Klenk H-D, Sanchez A (1993) Molecular biology and evolution of filoviruses. In: Kaaden OR, Eichhorn W, Czerny CP (eds) Unconventional agents and unclassified viruses. Recent advances in biology and epidemiology. Springer, Wien New York, pp 81–100 (Arch Virol [Suppl] 7)

23. Feldmann H, Nichol ST, Klenk H-D, Peters CJ, Sanchez A (1994) Characterization of filoviruses based on differences in structure and antigenicity of the virion glycoprotein. Virology 199: 469–473

24. Feldmann H, Sanchez A (1995) Detection of Marburg and Ebola virus infections by polymerase chain reaction assays. In: Becker Y, Darai G (eds) Frontiers in virology – diagnosis of human viruses by polymerase chain reaction technology. Springer, Berlin Heidelberg New York Tokyo, pp 411–418

25. Feldmann H, Will C, Schikore M, Slenczka W, Klenk H-D (1991) Glycosylation and oligomerization of the spike protein of Marburg virus. Virology 182: 353–356

26. Fisher-Hoch SP, Brammer L, Trappier SG, Hutwagner LC, Farrar BB, Ruo SL, Brown BG, Hermann LM, Perez-Oronoz GI, Goldsmith CS, Hanes MA, McCormick JB (1992) Pathogenic potential of filoviruses: role of geographic origin of primate host and virus strain. J Infect Dis 166: 753–763

27. Foberg U, Fryden A, Isaksson B, Jahrling PB, Johnson A, McKee K, Niklasson B, Normann B, Peters CJ, Bengtsson M (1991) Viral hemorrhagic fever in Sweden: experiences from management of a case. Scand J Infect Dis 23: 143–151

28. Francis DP, Smith DH, Highton RB, Simpson DIH, Lolik P, Deng IM, Gillo AL, Idrtis AD, El Tahir B (1978) Ebola fever in the Sudan, 1976: epidemiological aspects of the disease. In: Pattyn SR (ed) Ebola virus hemorrhagic fever, 1st ed. Elsevier/North-Holland, Amsterdam, pp 129–135

29. Gear JSS, Cassel GA, Gear AJ, Trappler B, Clausen L, Meyers AM, Kew MC, Bothwell TH, Sher R, Miller GB, Schneider J, Koornhoff HJ, Comperts ED, Isaäcson M, Gear JHS (1975) Outbreak of Marburg virus disease in Johannesburg. Br Med J 4: 489–493

30. Geisbert TW, Jahrling JB, Hanes MA, Zack PM (1992) Association of Ebola-related Reston virus particles and antigen with tissue lesions of monkeys imported to the United States. J Comp Pathol 106: 137–152

31. Geyer H, Will C, Feldmann H, Klenk H-D, Geyer R (1992) Carbohydrate structure of Marburg virus glycoprotein. Glycobiology 2: 299–312

32. Gonzales JP, Johnson ED, Josse R, Merlin M, Georges AJ, Abandja J, Danyod M, Delaporte E, Dupont A, Ghogomu A, Kouka-Bemba D, Madelon MC, Sima A, Meunier DMY (1989) Antibody prevalence against hemorrhagic fever viruses in randomized representative central African populations. Res Virol 140: 319–331

33. Hass R, Maass G (1971) Experimental infection of monkeys with the Marburg virus. In: Martini GA, Siegert R (eds) Marburg virus disease, 1st ed. Springer, New York, pp 136–143

34. Hayes CG, Burans JP, Ksiazek TG, DelRosario RA, Miranda MEG, Manaloto CR, Barrientos AB, Robles CG, Dayrit MM, Peters CJ (1992) Outbreak of fatal illness among captive macques in the Philippines caused by an Ebola-related filovirus. Am J Trop Med Hyg 46: 664–671

35. Henderson BE, Kissling RE, Williams MC, Kafuko GW, Martin M (1971) Epidemioliogical studies in Uganda relating to the 'Marburg' agent. In: Martini GA, Siegert R (eds) Marburg virus disease, 1st ed. Springer, New York, pp 166–176

36. Hennessen W (1971) Epidemiology of 'Marburg virus' disease. In: Martini GA, Siegert R (eds) Marburg virus disease, 1st ed. Springer, New York, pp 161–165

37. Heymann DL, Weisfeld JS, Webb PA, Johnson KM, Cairns T, Berquist H (1980) Ebola hemorrhagic fever: Tandala Zaire, 1977–78. J Infect Dis 142: 372–376

38. Holland JJ (ed) (1993) Genetic diversity of RNA viruses. Curr Top Microbiol Immunol 176: 1–226

39. ICTV (1991) The order Mononegavirales. Paramyxovirus Study Group of the Vertebrate Subcommittee. Arch Virol 117: 137–140

40. Ivanoff B, Duquesnoy P, Languillat G, Saluzzo JF, Georges A, Gonzales JP, McCormick JB (1982) Haemorrhagic fever in Gabon. 1. Incidence of Lassa, Ebola and Marburg virus in Haut-Ogooue. Trans Soc Trop Med Hyg 76: 719–720

41. Jahrling PB (1995) Filoviruses and arenaviruses. In: Murray PR, Baron EJ, Pfaller MA, Tenover FC, Yolken RM (eds) Manual of clinical microbiology. American Society for Microbiology, Washington, pp 1068–1081

42. Jahrling PB, Geisbert TW, Galgard DW, Johnson ED, Ksiazek TG, Hall WC, Peters CJ (1990) Preliminary report: isolation of Ebola virus from monkeys imported to USA. Lancet 335: 502–505

43. Johnson ED, Gonzales JP, Georges A (1993) Haemorrhagic fever virus activity in equatorial Africa: distribution and prevalence of filovirus reactive antibodies in the Central African Republic. Trans Soc Trop Med Hyg 87: 530–535

44. Johnson ED, Gonzales JP, Georges A (1993) Filovirus activity among selected ethnic groups inhabiting the tropical rain forest of equatorial Africa. Trans Soc Trop Med Hyg 87: 536–538

45. Johnson BK, Ocheng D, Oogo S, Gitau LG, Wambui C, Gichogo A, Libondo D, Tukei PM, Johnson ED (1986) Seasonal variation in antibodies against Ebola virus in Kenyan fever patients. Lancet 1: 1160

46. Johnson KM, Lang JV, Webb PA, Murphy FA (1977) Isolation and partial characterization of a new virus causing acute hemorrhagic fever in Zaire. Lancet 1: 569–571

47. Johnson KM, Scribner CL, McCormick JB (1981) Ecology of Ebola virus: a first clue? J Infect Dis 143: 749–751

48. Kissling RE, Robinson RQ, Murphy FA, Whitfield FG (1968) Agent of disease contracted from green monkeys. Science 160: 888–890

49. Kissling RE, Murphy FA, Henderson BE (1970) Marburg virus. Ann N Y Acad Sci 174: 932–945

50. Kiley MP, Bowen ETW, Eddy GA, Isaäcson M, Johnson KM, McCormick JB, Murphy FA, Pattyn SR, Peters D, Prozesky OW, Regnery RL, Simpson DIH, Slenczka W, Sureau P, van der Groen G, Webb PA, Wulff H (1982) Filoviridae: a taxonomic home for Marburg and Ebola viruses? Intervirology 18: 24–32

51. Kiley MP, Cox NJ, Elliott LH, Sanchez A, DeFries R, Buchmeier MJ, Richman DD, McCormick JB (1988) Physicochemical properties of Marburg virus: evidence for

three distinct virus strains and their relationship to Ebola virus. J Gen Virol 69: 1957–1967

52. Ksiazek TG, Rollin PE, Jahrling PB, Johnson E, Dalgard DW, Peters CJ (1992) Enzyme immunosorbent assay for Ebola virus antigens in tissues of infected primates. J Clin Microbiol 30: 947–950

53. LeGuenno B, Formentry P, Wyers M, Gounon P, Walker F, Boesch C (1995) Isolation and partial characterization of a new strain of Ebola virus. Lancet 345: 1271–1274

54. Martini GA (1971) Clinical syndrom. In: Martini GA, Siegert R (eds) Marburg virus disease, 1st ed. Springer, New York, pp 1–9

55. Martini GA, Knauff HG, Schmidt HA, Mayer G, Baltzer G (1968) Über eine bisher unbekannte, von Affen eingeschleppte Infektionskrankheit: Marburg-Virus-Krankheit. Dtsch Med Wochenschr 93: 559–571

56. Martini GA, Siegert R (1971) Marburg virus disease, 1st ed. Springer, New York, pp 1–230

57. Mathiot CC, Fontenille D, Georges AJ, Coulanges P (1989) Antibodies to haemorrhagic fever viruses in Madagascar. Trans Soc Trop Med Hyg 83: 407–409

58. Murphy FA, van der Groen G, Whitfield SG, Lange JV (1978) Ebola and Marburg virus morphology and taxonomy. In: Pattyn SR (ed) Ebola virus hemorrhagic fever, 1st ed. Elsevier/North-Holland, Amsterdam, pp 61–84

59. Pattyn SR (1978) Ebola virus hemorrhagic fever, 1st ed. Elsevier/North-Holland, Amsterdam, pp 1–436

60. Pattyn SR (1977) Isolation of Marburg-like virus from a case of hemorrhagic fever in Zaire. Lancet 1: 571–573

61. Peters D, Müller G, Slenczka W (1971) Morphology, development, and classification of Marburg virus. In: Martini GA, Siegert R (eds) Marburg virus disease, 1st ed. Springer, New York, pp 68–83

62. Peters CJ, Johnson ED, Jahrling PB, Ksiazek TG, Rollin PE, White J, Hall W, Trotter R, Jaax N (1993) Filoviruses. In: Morse SS (ed) Emerging viruses. Oxford University Press, Oxford, pp 159–175

63. Peters CJ, Johnson ED, McKee KT (1991) Filoviruses and management and viral hemorrhagic fevers. In: Belshe RB (ed) Textbook of human virology. Mosby Year Book, St. Louis, pp 699–712

64. Peters CJ, Sanchez A, Feldmann H, Rollin PE, Nichol ST, Ksiazek TG (1994) Filoviruses as emerging pathogens. Semin Virol 5: 147–154

65. Pokhodyaeu VA, Gonchar NI, Pshenichnov VA (1991) Experimental study of Marburg virus contact transmission. Vopr Virus 36: 506–508

66. Regnery RL, Johnson KM, Kiley MP (1980) Virion nucleic acid of Ebola virus. J Virol 36: 465–469

67. Richman DD, Cleveland PH, McCormick JB, Johnson KM (1983) Antigenic analysis of strains of Ebola viruses: identification of two Ebola virus subtypes. J Infect Dis 147: 268–271

68. Sanchez A, Kiley MP, Holloway BP, Auperin DD (1983) Sequence analysis of the Ebola virus genome: organization, genetic elements, and comparison with the genome of Marburg virus. Virus Res 29: 215–240

69. Sanchez A, Trappier S, Mahy BWJ, Peters CJ, Nichol ST (1996) The virion glycoproteins of Ebola viruses are encoded in two reading frames and are expressed through transcriptional editing. Proc Natl Acad Sci USA (in press)

70. Siegert R, Shu H-L, Slenczka W, Peters D, Müller G (1967) Zur Äthiologie einer unbekannten von Affen ausgegangenen Infektionskrankheit. Dtsch Med Wochenschr 92: 2341–2343

71. Slenczka W, Rietschel M, Hoffmann C, Sixl W (1984) Seroepidemiologische Untersuchungen über das Vorkommen von Antikörpern gegen Marburg- und Ebola-Virus in Afrika. Mitt Oesterr Ges Tropenmed Parasitol 6: 53–60

72. Smith CEG, Simpson DIH, Bowen ETW (1967) Fatal human disease from vervet monkeys. Lancet II: 1119–1121

73. Smith DH, Francis D, Simpson DIH, Highton RB (1978) The Nzara outbreak of haemorrhagic fever. In: Pattyn SR (ed) Ebola virus hemorrhagic fever, 1st ed. Elsevier/North-Holland, Amsterdam, pp 137–141

74. Smith DH, Johnson BK, Isaäcson M, Swanapoel R, Johnson KM, Kiley MP, Bagshawe A, Siongok T, Keruga WK (1982) Marburg-virus disease in Kenya. Lancet 1: 816–820

75. Stille W, Böhle E (1971) Clinical course and prognosis of Marburg virus ('green monkey') disease. In: Martini GA, Siegert R (eds) Marburg virus disease, 1st ed. Springer, Berlin Heidelberg New York, pp 10–18

76. Stille W, Böhle E, Helm E, vanRey W, Siede W (1968) Über eine durch Cercopithecus aethiops übertragene Infektionskrankheit. Dtsch Med Wochenschr 93: 572–582

77. Stojkovic LJ, Bordjoski M, Gligic A, Stefanovic Z (1971) Two cases of cercopithecus-monkeys-associated hemorrhagic fever. In: Martini GA, Siegert R (eds) Marburg virus disease, 1st ed. Springer, Berlin Heidelberg New York, pp 24–33

78. Swanepoel R (1987) Viral haemorrhagic fever in South Africa: history and national strategy. S Afr J Sci 83: 80–88

79. Teepe RGC, Johnson BK, Ocheng D, Gichogo A, Langatt A, Ngindu A, Kiley M, Johnson KM, McCormick JB (1983) A probable case of Ebola virus hemorrhagic fever in Kenya. East Afr Med J 60: 718–722

80. Tomori O, Fabiyi A, Sorungbe A (1988) Viral hemorrhagic fever antibodies in Nigerian population. Am J Trop Med Hyg 38: 407–410

81. van der Groen G, Johnson KM, Webb FA, Wulff H, Lange J (1978) Results of Ebola antibody survey in various population groups. In: Ebola virus hemorrhagic fever, 1st ed. Elsevier/North-Holland, Amsterdam, pp 203–208

82. van der Waals FJ, Pomerov KL, Goudsmit J, Asher DM, Gajdusek DC (1986) Hemorrhagic fever virus infection in an isolated rainforest area of central Liberia. Limitations of the indirect immunofluorescence slide test for antibody screening in Africa. Trop Geogr Med 38: 209–214

83. Volchkov VE, Blinov VM, Netesov SV (1992) The envelope glycoprotein of Ebola virus contains an immunosuppressive-like domain similar to oncogenic retroviruses. FEBS Lett 305: 181–184

84. Volchkov VE, Becker S, Volchkova VA, Ternovoj VA, Kotov AN, Netesov SV, Klenk H-D (1995) GP mRNA of Ebola virus is edited by the Ebola virus polymerase and by T7 and vaccinia virus polymerases. Virology 214: 421–430

85. World Health Organization (1978) Ebola hemorrhagic fever in Sudan, 1976. Bull World Health Organ 56: 247–270

86. World Health Organization (1978) Ebola hemorrhagic fever in Zaire, 1976. Bull World Health Organ 56: 271–293

87. World Health Organization (1979) Viral hemorrhagic fever surveillance. Wkly Epidemiol Rec 54: 342–343

88. World Health Organization (1985) Arthropod-borne and rodent-borne viral diseases. WHO Technical Report Series, no. 719

89. World Health Organization (1992) Viral hemorrhagic fever in imported monkeys. Wkly Epidemiol Rec 67: 142–143

90. World Health Organization (1995) Ebola hemorrhagic fever. Wkly Epidemiol Rec 70: 149–152
91. World Health Organization (1995) Ebola hemorrhagic fever. Wkly Epidemiol Rec 70: 241–242
92. World Health Organization (1995) Viral hemorrhagic fever – management of suspected cases. Wkly Epidemiol Rec 70: 249–256
93. Zaki SR (1995) Pathology of Ebola virus hemorrhagic fever. European Conference on Tropical Medicine, Hamburg, Germany: A22, p 3

Authors' address: Dr. H. Feldmann, Institute of Virology, Philipps- University, Robert-Koch Strasse. 17; D-35037 Marburg, Federal Republic of Germany.

Arch Virol (1996) [Suppl] 11: 101–114

Archives _of_ Virology
© Springer-Verlag 1996

# Characterization of a new Marburg virus isolated from a 1987 fatal case in Kenya

**E. D. Johnson**[1], **B. K. Johnson**[2], **D. Silverstein**[3], **P. Tukei**[2], **T. W. Geisbert**[1], **A. N. Sanchez**[4], and **P. B. Jahrling**[1]

[1] United States Army Medical Research Institute of Infectious Diseases, Fort Detrick, Frederick, Maryland, U.S.A., [2] Virus Research Center, Kenya Medical Research Institute, [3] The Nairobi Hospital, Nairobi, Kenya, [4] Special Pathogens Branch, Centers for Disease Control, Atlanta, Georgia, U.S.A.

**Summary.** In 1987, an isolated case of fatal Marburg disease was recognized during routine clinical haemorrhagic fever virus surveillance conducted in Kenya. This report describes the isolation and partial characterization of the new Marburg virus (strain Ravn) isolated from this case. The Ravn isolate was indistinguishable from reference Marburg virus strains by cross-neutralization testing. Virus particles and aggregates of Marburg nucleocapsid matrix in Ravn-infected vero cells, were visualized by immunoelectron microscopic techniques, and also in tissues obtained from the patient and from inoculated monkeys. The cell culture isolate produced a haemorrhagic disease typical of Marburg virus infection when inoculated into rhesus monkeys. Disease was characterized by the sudden appearance of fever and anorexia within 4 to 7 days, and death by day 11. Comparison of nucleotide sequences for portions of the glycoprotein genes of Marburg-Ravn were compared with Marburg reference strains Musoki (MUS) and Popp (POP). Nucleotide identity in this alignment between RAV and MUS is 72.3%, RAV and POP is 71%, and MUS and POP is 91.7%. Amino acid identity between RAV and MUS is 72%, RAV and POP is 67%, and MUS and POP is 93%. These data suggest that Ravn is another subtype of Marburg virus, analogous to the emerging picture of a spectrum of Ebola geographic isolates and subtypes.

## Introduction

The filovirus pathogens, Marburg and Ebola viruses, have been associated with dramatic outbreaks of viral haemorrhagic fever and high mortality rates in sporadic occurrences since Marburg virus was first recognized in 1967 [17, 18, 20]. The natural histories of these viruses remain obscure. Each occurrence of filovirus disease is viewed as an opportunity to identify the natural reservoir of filoviruses in nature, and representative isolates are characterized in comparison with reference strains for clues to viral maintenance and spread.

In 1987, an isolated case of fatal Marburg disease was recognized during routine clinical haemorrhagic fever virus surveillance conducted in Kenya by the Virus Research Centre at the Kenya Medical Research Institute (KEMRI), the Nairobi Hospital, and the United States Army Medical Research Institute of Infectious Diseases (USAMRIID). This report describes the isolation and partial characterization of a new Marburg virus (strain Ravn) isolated from this case. Until recently, the Ravn isolate was thought to be indistinguishable from reference strains associated with the first occurrence in Germany and Yugoslavia in 1967 and Kenya in 1980 [20]. However, modern techniques in molecular virology which have been applied to develop phylogenetic relation-ships [7] among the disparate species of Ebola virus (i.e. Ebola-Zaire, -Sudan, -Reston, and Ivory Coast) have also been applied to discriminate Marburg-Ravn from reference Marburg virus strains. Since this new virus subtype may prove to be epidemiologically important in some future outbreak, certain aspects of the informal epidemiological investigation which ensued are also reported.

## Materials and methods

### Virus isolation

Serum from blood collected during clinical disease was inoculated into 25 cm$^2$ tissue culture flasks containing confluent monolayers of Vero (African green monkey kidney cells), Vero E6 (a clone of Vero cells) and SW-13 (human adrenal cortex adenocarcinoma) cell cultures grown in Eagle minimum essential medium (EMEM) with 2 mmol L-glutamine, 10% heat inactivated fetal bovine serum, 100 U of penicillin, and 0.5 µg of streptomycin per ml. The cultures were incubated at 37 °C in a 5% CO$_2$ atmosphere and observed daily by inverted phase microscopy for cytopathic effect (CPE). Cultures were harvested and subcultured when CPE was observed. Supernatant fluids were clarified and stored at $-70$ °C. Cell monolayers were trypsinized and the single cell suspension pelleted by centrifugation at 1500 RPMs for 10 min. Cells were resuspended in isotonic saline and dispensed in 25 µl drops onto teflon templated microscope slides, air-dried, fixed overnight at $-70$ °C in acetone, gamma irradiated while on dry ice to inactivate residual infectious virus, and screened for cell-associated viral antigens by the immunofluorescent antibody test (IFAT) as previously described [12, 14] and below.

Infectious clinical material was handled within negative pressure flexible plastic isolators at the Virus Research Center, and subsequent viral isolation and characterization utilized the maximum biological containment (BSL-4) facilities at USAMRIID.

### Indirect IFAT reagents and assay

The indirect fluorescent antibody test (IFAT) utilized mouse ascitic fluid monoclonal and polyclonal antibody preparations with specificities for Marburg, Ebola, and a variety of viruses associated with haemorrhagic fever in Africa including Congo-Crimean hemorrhagic fever, dengue, Lassa, West Nile, and Yellow Fever. Sera from confirmed convalescent human Marburg and Ebola cases were also used. Inactivated reference spot slides of acetone fixed cells infected with the corresponding viruses were prepared, stored at $-70$ °C, and used in an indirect IFAT [12, 14] using appropriate dilutions of commercially obtained, fluorescein-conjugated goat anti species immunoglobulin.

### Neutralization testing and viral assay

Infectious virus was assayed by counting plaque forming units (PFU) on vero cells maintained under agarose in $10 \, \text{cm}^2$ well of plastic tissue culture plates as described [12]. For measuring neutralizing (nt) antibody to filoviruses, test sera were diluted 1:10 in EMEM with 10% guinea pig serum as a complement source and mixed with serial dilutions of challenge virus, as previously detailed [12]. Titers are expressed as a $\log_{10}$ neutralization index (LNI), defined as $(\log_{10} \text{PFU in control}) - (\log_{10} \text{PFU in test serum})$. Infectious virus was assayed by counting plaque forming units (PFU) on monolayers of MA-104 cells maintained under agarose as previously described [12].

### Electron microscopy of cell pellets

Cells from $25 \, \text{cm}^2$ flasks were scraped into 1.5 ml of EMEM and centrifuged at 12 000 rpm in a conical tube for 30 sec to form a loose pellet. These, and autopsy tissues were immersion-fixed in either 2% glutaraldehyde in 0.1 M Millonig's phosphate buffer for standard transmission electron microscopy (TEM) or 2% paraformaldehyde $+ 0.1\%$ glutaraldehyde for immunoelectron microscopy (IEM), as described previously [11].

### Polymerase chain reaction and sequencing

The Musoke (MUS) and Popp (POP) strains of MBG were obtained from GenBank (accession numbers Z12132 and Z29337, respectively). The Ravn (RAV) strain sequence was derived from a reverse transcriptase-polymerase chain reaction (RT-PCR) product. Genomic RNA (vRNA) was extracted from virions, purified from clarified tissue culture supernatant fluids by pelleting through a 20% sucrose cushion by the method of Chomczyski and Sacchi [4]. The vRNA was used as template for the RT-PCR using a pair of MUS GP-specific oligodeoxynucleotide "primers", and the product was directly sequenced using an automated sequencer (ABI) and the same primers. Sequences used in alignments encode or correspond to amino acid sequences 175 to 274 of the GP. Alignments were produced using the PileUp computer program (Genetics Computer Group, Wisconsin Package, Version 8.0-OpenVMS AXP). Comparisons used "Gapweight" and "Gaplengthweight" settings of 5.0 and 1.0, respectively, for nucleotide sequences and 3.0 and 1.0 for amino acid sequences.

### Monkey inoculation

Six adult rhesus monkeys, *Macaca mulatta*, were screened and found negative for filovirus IFAT antibodies, then separated into two groups of three animals and maintained in individual cages on standard diets. Monkeys were inoculated intramuscularly (IM) with 0.5 ml of clarified and diluted (1:100) cell culture supernatants (sn) containing 10 000 PFU of infectious virus or control (uninfected sn). All animals were observed daily for signs of disease. On day 0 pre-exposure and days 4, 7 and 10 after exposure, animals were sedated with ketamine (5 mg/kg dose), and venous blood drawn for laboratory evaluation. Necropsies were performed on moribund euthanized animals and organ samples taken for virological and pathological studies.

Whole blood samples were anti-coagulated in 7.5% potassium EDTA for hematologic examination or in 3.8% sodium citrate for coagulation studies or fibrinogen level determinations. For fibrin-fibrinogen degradation product (FDP) measurements, whole blood was collected in glass tubes containing 10 NIH units of thrombin and 18 mg of EACA, and the serum removed after clotting. Serum for enzyme levels was separated from whole blood drawn into glass vacutainer tubes and allowed to clot. Total and differential white blood cell counts, hematocrit, and platelet counts were determined using a ELT-9/ds (Ortho Diagnostics, Westwood, MA) laser-based hematologic analyzer. Hemostasis was evaluated by determining activated partial thromboplastin times (APTT) on a MLA 750 coagulometer (Medical Laboratory Automation, Inc., Mount Vernon, NY) using standard assay procedures [21]. Fibrinogen was measured by the method of Clauss and FDP levels by the staphylococal clumping assay [5, 11]. Serum aspartate aminotransferase (AST) and lactic dehydrogenase (LDH) measurements were performed using a Multistat centrifugal chemical analyzer (Instrument Laboratories, Lexington, MA).

### Epidemiological investigation

An informal medical field team was assembled, shortly after the laboratory diagnosis of Marburg virus disease was obtained, to conduct an epidemiological investigation of the incident. The patient's travels during the 21 days prior to disease onset were reconstructed, and likely locations where the disease could have been contracted were visited. Local hospitals and clinics were visited to identify unusual morbidity or mortality and review likely cases. The general health and serological status of those exposed to the patient prior to and during his illness were also evaluated. Venous blood was drawn aseptically according to accepted procedures.

## Results

### Case history

On 13 August 1987, a 15 year old European male who had been in Kenya for 1 month, was admitted to the Aga Khan Hospital, Mombassa, with a 3 day history of headache, malaise, anorexia, fever and vomiting. He had been taking chloroquine phosphate prophylactically and prior to admission had been treated with Fansidar (pyrimethamine sulfadoxine) and Camoquin (amodiaquine). On admission, a blood film was malaria parasite negative, but since no history of amodiaquine treatment was given, he was treated with amodiaquine and Metakalfin (sulfamethopyrazine). The white blood cell count was 1 800 per mm$^3$. The patient steadily deteriorated, developing bloody diarrhea which became copious over a 3 day period, and hypotension. His white blood cell count rose to 20 000 per mm$^3$. He was transferred to the intensive care unit of the Aga Khan Hospital on the eighth day of illness. On the ninth day, he was transferred to the Nairobi Hospital by the Flying Doctors Service. On admission to the Nairobi Hospital, the patient was febrile (40 °C) and delirious. Blood pressure was 80/60 and subsequently fell to unrecordable levels. The patient was cyanotic and had cutaneous lesions resembling urticaria, which later became dark blue patches. Laboratory tests revealed a thrombocytopenia (26 000 per mm$^3$) and prolonged prothrombin and partial thromboplastin times. The Widal test was negative. Despite intensive supportive therapy, high dose steroids, chloramphenicol and Cloxacillin, heparin, fresh plasma, blood, and haemoceal, the patient became haemorrhagic and toxic. The patient continued to be hypotensive, developed elevated blood potassium and urea levels, and was dialyzed twice but continued to deteriorate. He developed dark blue cutaneous lesions which appeared to be small microthrombi and part of a disseminated intravascular coagulation (DIC) process, and died of cardiac failure the morning of the 11th day of illness. The provisional diagnoses were viral haemorrhagic fever, most likely Marburg virus disease based upon the characteristic clinical picture [9, 20], or less likely, typhoid fever with extreme toxicity [15, 18].

Post mortem examination revealed a significant haemorrhagic disease with massive petechial and purpuric hemorrhage in the skin, conjunctiva, and gastrointestinal mucosa. Copious blood tinged pleural, pericardial, and peritoneal effusions were present as well as marked retropertioneal edema. The lungs and tracheobronchial tree were hemorrhagic. Multiple hemorrhagic petechiae were observed on the epicardium, renal cortices and pelvises and bladder. No haemorrhage was observed in the acutely congested liver, spleen, pancreas, or adrenals. Histological examination revealed 3 pathological processes: diffuse coagulative necrosis resulting from clinical shock, micro-thrombi with surrounding tissue infarction resulting from DIC, and micro-abscesses containing Gram negative rod-shaped microorganisms.

*Pseudomonas aeruginosa* was isolated from blood, tracheal aspirate, and urine cultures of samples drawn at admission to the Nairobi Hospital. The gross pathological findings suggested acute cardiac decompensation as a result of septicemia and the preliminary anatomic diagnosis suggested that bacteremic shock was the immediate cause of death.

### Isolation and characterization of the virus

Acute serum collected on the ninth day of illness was found to contain infectious Marburg virus by cell culture inoculation. Vero E6 cell cultures inoculated with the patient serum developed signs of cytopathic granulation and rounding by 5 days post-inoculation (PI). Acetone-fixed Vero E6 cells from the primary culture reacted positively with the patient's serum in the indirect IFAT. SW-13 cells inoculated with the patient's acute serum or passaged Vero E6 supernatant fluid reacted positively on day 5 PI with the patient's serum, reference MBG virus-immune human convalescent plasma and anti-MBG mouse monoclonal antibodies, but did not react with antisera against Ebola virus or Rift Valley fever virus by IFAT (Table 1). CPE was observed in SW-13 cell culture on day 9 PI. Fixed SW-13 cells from CPE positive cultures did not react with antibodies against Congo-Crimean hemorrhagic fever, Lassa, dengue, West Nile or yellow fever viruses by indirect IFAT. The patient's day 9 serum reacted at a dilution of 1:16 against antigens produced in Vero cells infected with the reference Musoke strain of MBG virus, recovered during the 1980 MBG disease episodes.

Cross-neutralization testing confirmed that the virus isolated was closely related to the reference (Musoki) Marburg strain (Table 2). Serum from guinea pigs (GP) surviving Ravn strain neutralized 2.3 $\log_{10}$ PFU of the Ravn virus and 2.1 $\log_{10}$ PFU of the Musoki strain but failed to neutralize Ebola-Reston strain. Likewise, GP anti-Musoki neutralized Ravn and Musoki equally but failed to neutralize Ebola-Reston, as did monkey anti-Ravn serum. Conversely, both GP

**Table 1.** IFAT reactivity of acetone-fixed SW-13 cells incubated with the suspected filovirus isolate[a]

| Diagnostic reagents | | | IFA reactivity: inoculum | | | |
| | Virus specificity | Dilution | Patient isolate | Virus[b] | | |
| Specimens | | | | MBG | EBO | RVF |
| --- | --- | --- | --- | --- | --- | --- |
| Acute serum[c] | ? | 1:16 | 1+ | 2+ | − | − |
| Reference immune serum | MBG | 1:16 | 3+ | 3+ | − | − |
| | EBO | 1:16 | − | − | 3+ | − |
| Non-immune human serum | − | 1:16 | − | − | − | − |
| Monoclonal antibody | MBG (CB01-BB01) | 1:100 | 3+ | 3+ | − | − |
| | | 1:1000 | + | 1+ | − | − |
| | EBO (DD4AE-8A1) | 1:100 | − | − | 3+ | − |
| | | 1:1000 | − | − | 3+ | − |
| | RVF (R3-1D8-1-1A) | 1:1000 | − | − | − | 3+ |

[a] IFAT staining intensity was ranked (− negative, + trace, 1+ weak, 2+ moderate, 3+ strong fluorescence) in comparison with reference reagents
[b] Abbreviations: *MBG* Marburg virus; *EBO* Ebola virus; *RVF* Rift Valley fever virus
[c] The acute serum specimen was collected nine days after onset

**Table 2.** Cross-neutralization test with Marburg-Ravn and reference Marburg and Ebola virus strains

| Specificity of antiserum | | LNI[a] of antiserum against virus | | |
| | | Marburg-Ravn | Marburg-Musoki | Ebola-Reston |
| --- | --- | --- | --- | --- |
| *Species* | *Virus strain* | | | |
| GP | Mbg-Ravn | 2.3 | 2.1 | 0.0 |
| GP | Mbg-Musoki | 1.9 | 2.0 | 0.1 |
| GP | Ebola-Reston | 0.0 | 0.0 | 2.5 |
| Monkey | Mbg-Ravn | 1.7 | 1.6 | 0.1 |
| Monkey | Ebola-Reston | 0.0 | 0.0 | 2.5 |

[a] LNI = $Log_{10}$ neutralization index of serum diluted 1:10

and monkey anti-Ebola-Reston sera neutralized Ebola-Reston but not Ravn or Musoki.

The serologic diagnosis of Marburg virus was corroborated by electron microscopic examination of tissues taken at autopsy from the patient. Although tissue preservation was not optimal, virus particles morphologically resembling

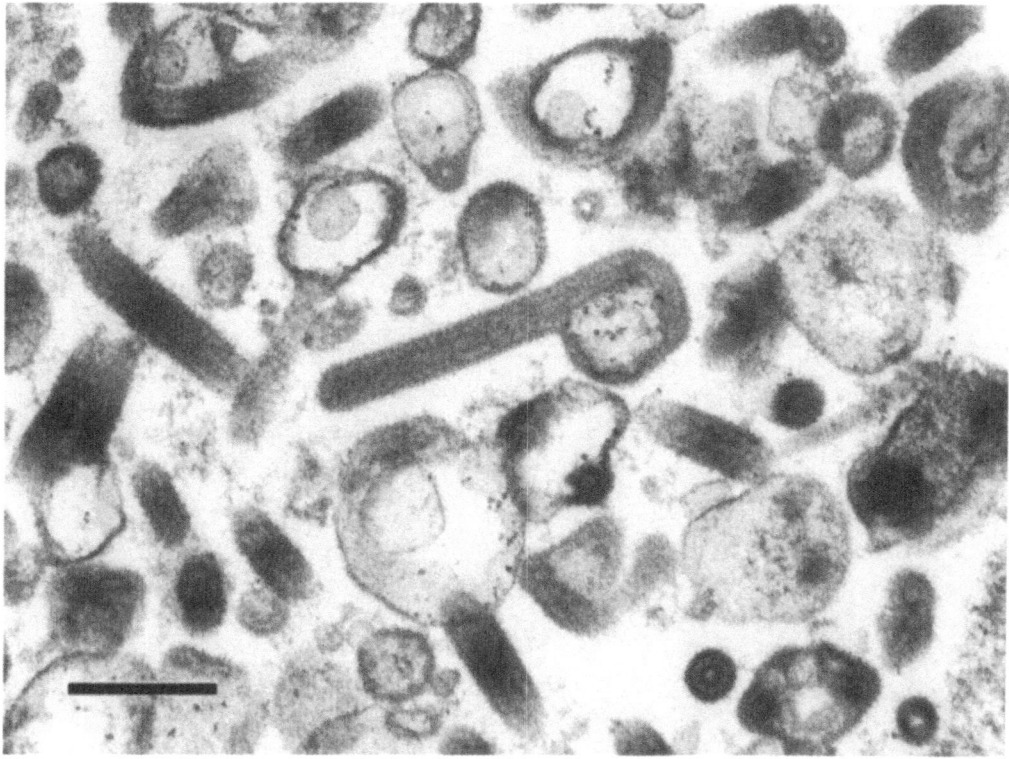

**Fig. 1.** Thin section through human adrenal cortex showing filovirus-like particles frequently seen in intercellular spaces. Bar: 250 nm

characteristic filovirions [8] were clearly visualized in some tissues such as adrenal (Fig. 1). Immunoelectron microscopy of MBG-infected Vero cells showed positive gold-sphere labeling of virus particles and intracytoplasmic viral inclusions when incubated with a monoclonal antibody specific for MBG nucleoprotein (NP) (Fig. 2).

### Genetic characterization of Marburg-Ravn

The Ravn (RAV) strain sequence for a region of the glycoprotein gene was derived from a reverse transcriptase-polymerase chain reaction (RT-PCR) product and compared with the sequences of two reference strains of Marburg, Musoke (MUS) and Popp (POP) (Fig. 3). Nucleotide (A) and predicted amino acid (B) sequences for three strains of MBG are shown with consensus sequences at the bottom of the alignments. Sequences used in alignments encode or correspond to amino acid sequences 175 to 274 of the GP. Sequences shown in lanes of MBG strains differ with the consensus (2 or more identical bases/residues). Nucleotide identity in this alignment between RAV and MUS is 72.3%, RAV and POP is 71%, and MUS and POP is 91.7%. Amino acid identity between RAV and MUS is 72%, RAV and POP is 67%, and MUS and POP is 93%.

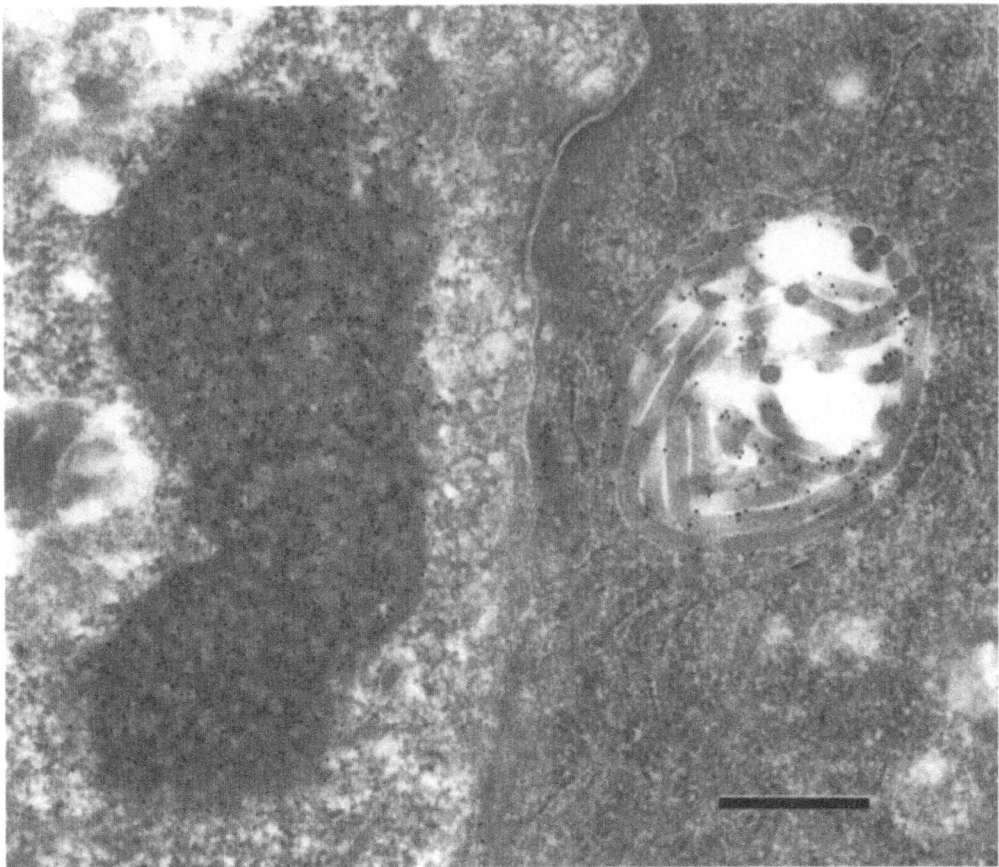

**Fig. 2.** Thin section through Ravn-infected Vero cells showing positive gold-sphere labeling of filovirus particles and an intracellular filoviral inclusion when incubated with a murine monoclonal antibody directed against a MBG viral nucleoprotein. Bar: 385 nm

### *Experimental infection of rhesus monkeys with Marburg-Ravn strain*

The cell culture isolate produced a hemorrhagic disease typical of Marburg virus infection [4, 19] when inoculated into rhesus monkeys. The onset of disease, between 4 and 7 days post-inoculation was sudden and accompanied by fever and anorexia. A petechial rash and hemorrhagic diathesis developed. Elevated coagulation times, hypofibrinogenemia, and increased levels of circulating fibrin degradation products were also observed (Table 3). The disease, however, did not prove uniformly fatal; one of three ill animals, 8A40, started to improve on day 11 and made a full recovery, eventually seroconverting by both the IFAT and neutralizing antibody assays. A detailed description of the anatomical lesions in the two lethally infected monkeys will be reported separately, (Jaax et al., in prep.). In brief, histopathological examination of tissues from the two fatal rhesus infections revealed hemorrhage in all major organ systems. Filovirus particles were observed by EM in necropsy samples of liver, lung, spleen, lymph nodes, kidney, and adrenal samples (Fig. 3).

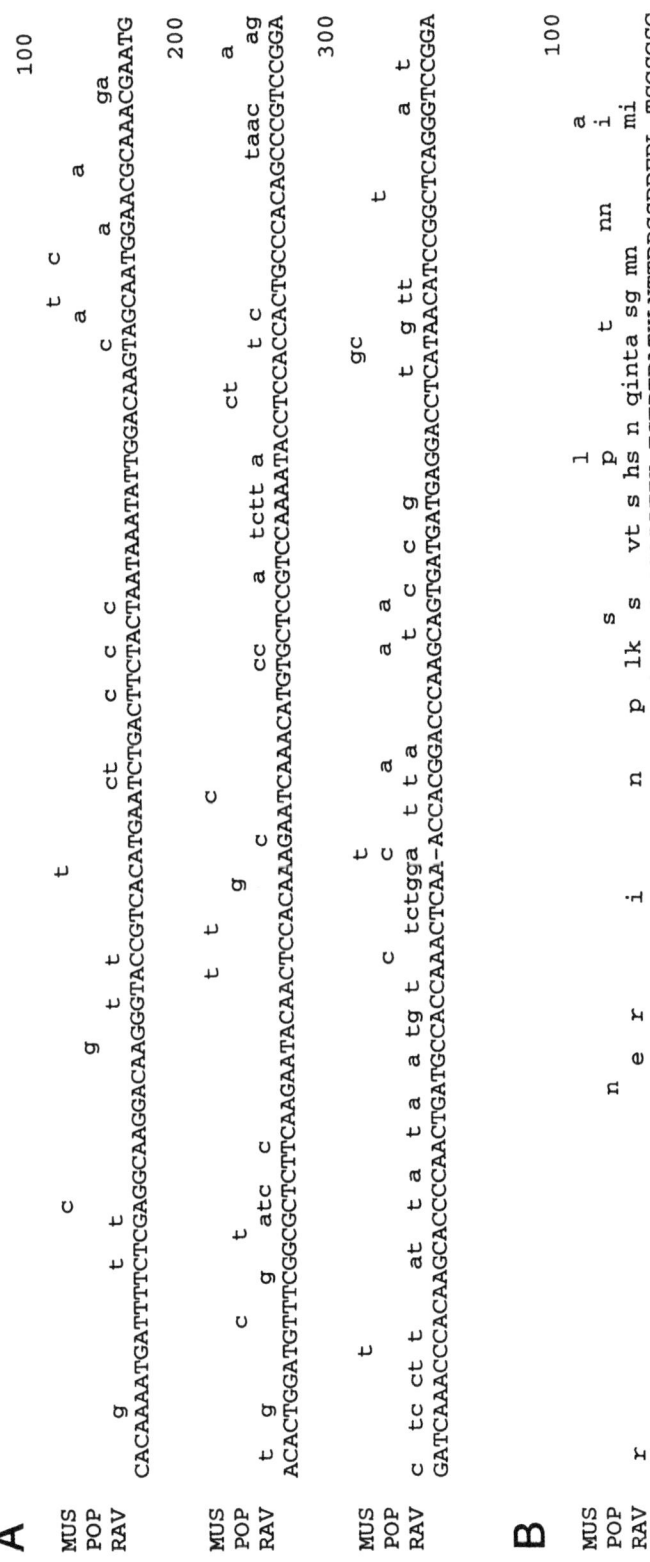

**Fig. 3.** Alignments of Marburg virus (MBG) glycoprotein (GP) gene sequences. Nucleotide (**A**) and predicted amino acid (**B**) sequences for three strains of MBG are shown with consensus sequences at the bottom of the alignments. The Musoke (*MUS*) and Popp (*POP*) strains of MBG were obtained from GenBank (accession numbers Z12132 and Z29337, respectively). The Ravn (*RAV*) strain sequence was derived from a reverse transcriptase-polymerase chain reaction (RT-PCR) product. Purified genomic RNA was used as template for the RT-PCR using a pair of MUS GP-specific oligodeoxynucleotide "primers", and the product was directly sequenced using an automated sequencer (ABI) and the same primers. Sequences used in alignments encode or correspond to amino acid sequences 175 to 274 of the GP. Alignments were produced using the PileUp computer program (Genetics Computer Group, Wisconsin Package, Version 8.0-OpenVMS AXP). Comparisons used ~Gapweight" and UGaplengthweight" settings of 5.0 and 1.0, respectively, for nucleotide sequences and 3.0 and 1.0 for amino acid sequences. Sequences shown in lanes of MBG strains differ with the consensus (2 or more identical bases/residues). Nucleotide identity in this alignment between RAV and MUS is 72.3%, RAV and POP is 71%, and MUS and POP is 91.7%. Amino acid identity between RAV and MUS is 72%, RAV and POP is 67%, and MUS and POP is 93%

**Table 3.** Hemorrhagic diathesis in rhesus monkeys inoculated with the Marburg-Ravn isolate[a]

| Animal number | Exposure days[b] | Platelets ($\times 10^3/mm^3$) | AST (IU/L) | LDH (IU/L) | APTT (SEC) | Fibrinogen (mg/dl) | FDP (µg/ml) |
|---|---|---|---|---|---|---|---|
| | | Clinical parameters | | | | | |
| P986 | 0 | 285 | 35 | 167 | 21.3 | 248 | 0.1 |
| | 7 | 89 | 473 | 4680 | 48.7 | 103 | 102.4 |
| 18295 | 0 | 281 | 112 | 366 | 21.1 | 340 | 0.4 |
| | 7 | 179 | 4706 | 15990 | 87.3 | 137 | 102.4 |
| 8A40[c] | 0 | 246 | 40 | 232 | 21.0 | 250 | 1.6 |
| | 7 | 183 | 368 | 814 | 25.9 | 350 | 12.8 |
| Controls[d] | 0 | 319 ± 38 | 44 ± 8 | 190 ± 45 | 22.6 ± 1 | 236 ± 126 | 0.3 ± 0.1 |
| | 7 | 365 ± 35 | 67 ± 37 | 262 ± 61 | 23.7 ± 3 | 215 ± 72 | 0.3 ± 0.2 |

[a] Animals P986, 18295 and 8A40 were inoculated IM with 0.5 mls of clarified infected SW-13 cell culture fluids, 10,000 PFU

[b] Clinical parameters were measured on day 0-pre-exposure and day 7 after exposure

[c] Monkey 8A40 survived virus challenge and developed Marburg virus-reactive IFAT and neutralizing antibody

[d] Mean ± SD of clinical parameters measured in three control animals inoculated IM with 0.5 ml of clarified uninfected SW-13 cell culture fluid

*Epidemiological studies*

The patient was a 15 year old boy who had arrived in Kenya from Europe on 14 July, 1987. Detailed epidemiological studies tracing the movements and contacts of the patients will be reported separately. He had spent most of the first two weeks in Kisumu, an urban setting on the shore of Lake Victoria in Nyanza Province. The patient's stay in Kisumu was uneventful. On 2 August (9 days prior to onset) the family visited Kitum Cave in Mount Elgon National Park; the patient spent at least 45 min exploring the cave. Kitum Cave is very large; it is frequented by many types of feral animals including elephants, buffalo, monkey, predator cats, birds, bats, and small rodents species. During his visit, the young man penetrated to the depth of the cave, about 500 meters from the mouth, and also took several trips to side chambers within the cave to collect crystals unique to Kitum. He may have received small cuts or puncture wounds from the crystals which were needle sharp. The family returned to Kisumu after the cave expedition. On 8 August, the family traveled by road from Kisumu to Mombasa, arriving on 9 August. He began to feel ill on 11 August, and was admitted to Aga Khan Hospital two days later.

A number of medical workers were exposed to the patient during the course of his illness, in addition to family members and domestic staff. A total of 64 known contacts of the patient were bled between 9 September and 16 October, 24 days or more after last possible exposure to the patient. All were found negative for MBG virus antibodies except one doctor from the Nairobi Hospital who had suffered a virologically and serologically confirmed and serious clinical case of MBG virus disease in 1980 and was known at the time of our survey to have pre-existing antibodies.

## Discussion

Marburg virus may continue to cause sporadic cases of fatal disease in Africa. The latest documented case occurred in August 1987, as reported here. The illness was suspected from the clinical picture which was similar to previously described cases of Marburg virus disease [6, 9, 20] but complicated in the terminal stages, by a *Pseudomonas aeruginosa* septicemia. Marburg virus was isolated in cell culture from an acute blood sample, and was recognized as a filovirus by electron microscopy of infected cell cultures and post-mortem tissue samples. The virus isolate was demonstrated to induce a haemorrhagic disease typical of Marburg virus infection in rhesus monkeys. The clinical diagnosis of Marburg virus disease and successful virological confirmation of the diagnosis in 1987 was facilitated by the proximity of a well equipped hospital with staff who recognized Marburg virus disease, and a virology laboratory with specialized expertise and containment capability to manage the highly pathogenic African hemorrhagic fever viruses.

Since this viral isolation was accomplished in 1987, the virological tools available for diagnosis of Marburg have improved substantially. An antigen capture ELISA has been developed which facilitates the detection of antigenemia

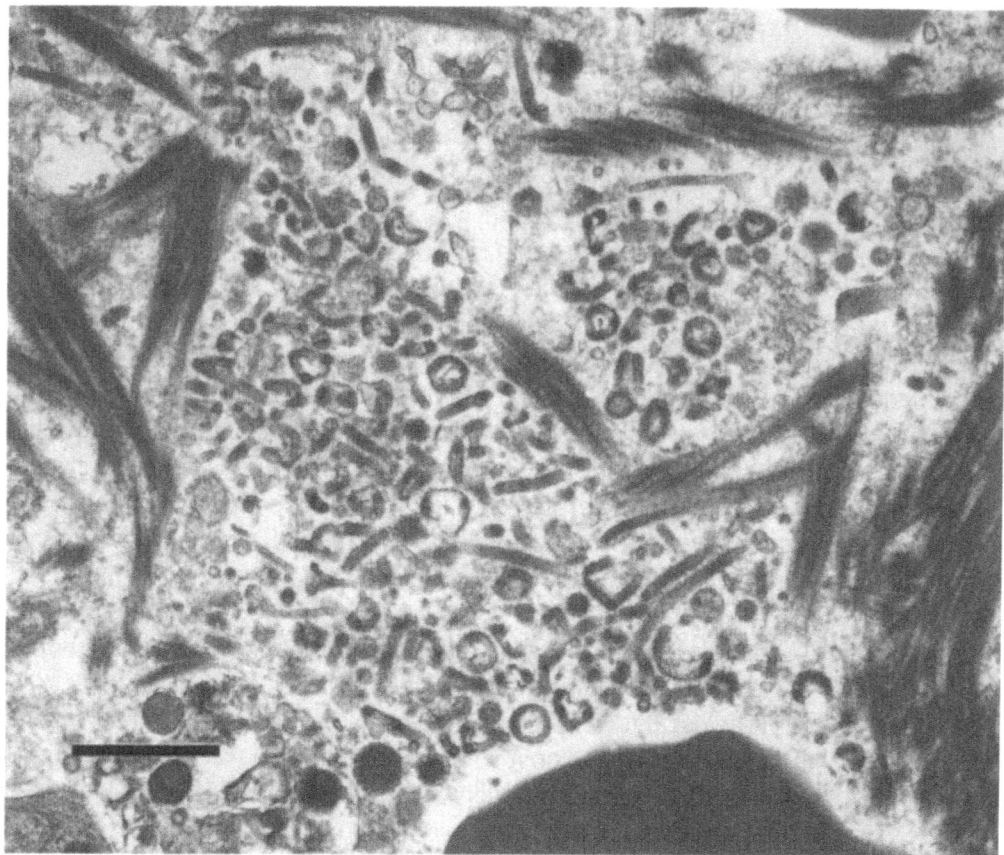

**Fig. 4.** Thin section through spleen of monkey experimentally infected with MBG-Ravn virus showing aggregate of filovirus particles often seen in areas of fibrin deposition. Bar: 930 nm

in serum [12]; this technique has been successfully applied to formalin-inactivated specimens, which should reduce the biohazard of the procedure. Application of polymerase chain reaction has not been tested in a clinical setting with Marburg, but retrospective analysis of specimens banked from the original 1967 outbreak in Germany and Yugoslavia could confirm its potential. Certainly, PCR has proven to be useful in achieving a timely diagnosis of other filovirus outbreaks, including the 1995 Ebola outbreak in Zaire [2]. One potential shortfall in the PCR technique is its strict specificity; consensus primers must be employed to ensure that variant strains, analogous to the Ebola-Reston [13] and -Ivory Coast strains [16] are not missed. There will always be a requirement to isolate and characterize the virus however, and for this, biocontainment facilities are a necessity. The lack of secondary cases was fortuitous; there was ample opportunity for secondary infection to occur during the incident. The patient's lungs, tracheobronchial tree, gastrointestinal tract and bladder were all severely hemorrhagic. During his illness the patient had frequent contact with health care workers and family members. The patient experienced severe vomiting and bloody diarrhea and required numerous bedding changes. The patient was flown from Mombasa to Nairobi in a small aircraft and after admission to Nairobi Hospital was dialyzed twice. The lack of secondary cases, however, could also be a reflection of potentially lower virulence for this strain, a testable hypothesis in monkeys. Recommendations for the safe management of viral hemorrhagic fever patients have been recently updated [3]. Genetic comparisons suggest this strain (Ravn) is clearly distinct from reference strains isolated from this same geographic area of East Africa. This genetic variation may have significance in both epidemiology and pathogenicity of Marburg virus strains.

Despite the genetic disparity of the isolates from the 1980 and 1987 episodes, the similarity of circumstances for the two exposures is striking. Marburg virus activity seems to occur, at least periodically, in the area between Lake Kyoga (Uganda) and Mount Elgon (Kenya). In this area, humans are incidental hosts who become involved through accidental contact with a reservoir species [1]. The fact that the 1987 patient and the 1980 Marburg virus disease index case had visited Mount Elgon National Park and entered Kitum cave 9 and 14 days before falling ill is intriguing and suggests a possible site for the accidental interaction between infected animals and man. However, an ecological survey conducted at Kitum Cave shortly after the 1987 exposure failed to identify the source of the infection or the mode of transmission. Reassessment of the numerous specimens from this survey using the PCR assay is feasible and potentially important. A breakthrough in understanding the natural history of any of the filovirus pathogens might lend valuable insight into maintenance cycles of the others, including the Ebola strain which recently emerged in Zaire in 1995. Until we reach this understanding, we will remain unable to predict which changes in social practices, ecological factors, and genetic characteristics of the filoviruses will combine to result in the emergence of the next filovirus outbreak.

## Acknowledgements

We thank the Kenya Office of the President, Kenya Ministry of Health, Kenya Ministry of Wildlife Conservation and Tourism, Director of Wildlife Management, Director of the Institute of Primate Research and Director of the Kenya Medical Research Institute, and Dr. C. J. Peters for their enthusiastic support of the Marburg virus disease investigation.

## References

1. Arata AA, Johnson BK (1978) Approaches towards studies of potential reservoirs of viral haemorrhagic fever in Southern Sudan (1977). In: Pattyn SR (ed) Ebola virus hemorrhagic fever. Elsevier/North-Holland Biomedical Press, Amsterdam, pp 191–200
2. Centers for Disease Control and Prevention (1995) Update: outbreak of Ebola viral haemorrhagic fever-Zaire, 1995. MMWR 44: 381–382
3. Centers for Disease Control (1995) Management of patients with suspected viral haemorrhagic fever. MMWR 44: 475–479

4. Chomczynski P, Sacchi N (1987) Single step method of RNA isolation by acid guanidium thiocyanate-phenol-chloroform extraction. Anal Biochem 162: 156–159
5. Clauss A (1957) Rapid physiological coagulation method in determination of fibrinogen. Acta Haematol 17: 237–246
6. Conrad JL, Isaacson M, Smith EB, Wulff H, Grees M, Gildenhags P, Johnson J (1978) Epidemiologic investigation of Marburg virus disease South Africa, 1975. Am J Trop Med Hyg 27: 1210–1215
7. Feldmann H, Klenk HD, Sanchez A (1993) Molecular biology and evolution of filoviruses. In: Kaaden OR, Eichhorn W, Czerny CP (eds) Unconventional agents and unclassified viruses. Recent advances in biology and epidemiology. Springer, Wien New York, pp 81–100 (Arch Virol [Suppl] 7)
8. Geisbert TW, Jahrling PB (1995) Differentiation of filoviruses by electron microscopy. Virus Res 39: 129–150
9. Gear IS, Cassel GA, Geor AJ, Troppler B, Clausen L, Meyers AM, Kew MC, Bothwell TH, Sher R, Miller GB, Schneider J, Koomhof HJ, Gompets ED, Isaacson M, Geor JHS (1975) Outbreak of Marburg virus disease in Johannesburg. Br Med J 4: 489–493
10. Hawiger J (1980) Measurement of fibrinogen and fibrin degradation products in serum by staphylococcal clumping test. J Lab Clin Med 75: 93–108
11. Jahrling PB, Geisbert TW, Jaax NK, Hanes MA, Ksiazek TG, Peters CJ (1996) Experimental infection of cynomolgus macaques with Ebola-Reston filoviruses from the 1989–1990 U.S. epizootic. In: Schwarz TF, Siegl G (eds) Imported virus infections. Springer, Wien New York, pp 115–134 (Arch Virol [Suppl] 11)
12. Jahrling PB (1995) Filoviruses and arenaviruses. In: Murray PR, Baron EJ, Pfaller MA, Tenover FC, Yolken RH (eds) Manual of clinical microbiology, 6th ed. ASM Press, Washington, pp 1068–1081
13. Jahrling PB, Geisbert TW, Dalgard DW, Johnson ED, Ksiazek TG, Hall WC, Peters CJ (1990) Preliminary report: isolation of Ebola virus from monkeys imported to the USA. Lancet 335: 502–505
14. Johnson KM, Elliott LH, Heymann DL (1981) Preparation of polyvalent viral immuno-fluorescent intracellular antigens and use in human serosurveys. J Clin Microbiol 14: 527–529
15. Martini GA, Knauff HG, Schmidt HA, Mayer G, Baetzer G (1968) A hitherto unknown infectious disease contracted from monkeys. Marburg virus disease. Ger Med Mon 13: 457–470
16. Le Guenno B, Formentry P, Wyers M, Gounon P, Walker F, Boesch C (1995) Isolation and partial characterization of a new strain of Ebola virus. Lancet 345: 1271–1274
17. Peters CJ, Sanchez A, Feldmann A, Rollin PE, Nichol S, Ksiazek TG (1994) Filoviruses as emerging pathogens. Semin Virol 5: 147–154
18. Siegert R (1970) The Marburg virus (vervet monkey agent). Mod Trends Med Virol 32: 204–240
19. Simpson DIH, Zlotnbik I, Rutter DA (1968) Vervet monkey disease. Experimental infection of guinea pigs and monkeys with the causative agent. Br J Exp Pathol 69: 458–464
20. Smith DH, Johnson BK, Isaacson M, Swanepoel R, Johnson KM, Kiley MP, Bagshawe A, Siangok T, Keruga W (1982) Marburg virus disease in Kenya. Lancet i: 816–820
21. Triplett DA, Smith C (1982) Routine testing in the coagulation laboratory. In: Triplett DA (ed) Laboratory evaluation of coagulation. American Society of Clinical Pathologists Press, Chicago, pp 27–52

Authors' address: Dr. P. B. Jahrling, Scientific Advisor, Headquarters, USAMRIID, Fort Detrick, Frederick, MD 21702-5011, U.S.A.

Arch Virol (1996) [Suppl] 11: 115–134

# Experimental infection of cynomolgus macaques with Ebola-Reston filoviruses from the 1989–1990 U.S. epizootic

P. B. Jahrling, T. W. Geisbert, N. K. Jaax, M. A. Hanes,
T. G. Ksiazek*, and C. J. Peters*

United States Army Research Institute of Infectious Diseases, Fort Detrick,
Frederick, Maryland, U.S.A.

**Summary.** This study describes the pathogenesis of the Ebola-Reston (EBO-R) subtype of Ebola virus for experimentally infected cynomolgus monkeys. The disease course of EBO-R in macaques was very similar to human disease and to experimental diseases in macaques following EBO-Zaire and EBO-Sudan infections. Cynomolgus monkeys infected with EBO-R in this experiment developed anorexia, occasional nasal discharge, and splenomegaly, petechial facial hemorrhages and severe subcutaneous hemorrhages in venipuncture sites, similar to human Ebola fever. Five of the six EBO-R infected monkeys died, 8 to 14 days after inoculation. One survived and developed high titered neutralizing antibodies specific for EBO-R. The five acutely ill monkeys shed infectious virus in various bodily secretions. Further, abundant virus was visualized in alveolar interstitial cells and free in the alveoli suggesting the potential for generating infectious aerosols. Thus, taking precautions against aerosol exposures to filovirus infected primates, including humans, seems prudent. This experiment demonstrated that EBO-R was lethal for macaques and was capable of initiating and sustaining the monkey epizootic. Further investigation of this animal model should facilitate development of effective immunization, treatment, and control strategies for Ebola hemorrhagic fever.

## Introduction

The recent re-emergence of Ebola virus in Zaire [2, 3, 28] is but the last of a succession of filovirus [20] outbreaks [26] which have occurred sporadically since Marburg hemorrhagic fever was first recognized in 1967 [23]. In 1989, a previously unrecognized Ebola virus subtype (Ebola-Reston) was associated with an epizootic in cynomolgus monkeys imported into the United States from the Philippines [17]. The source of this virus was eventually traced back to

---

* Present address: Special Pathogens Branch, Centers for Disease Control, Atlanta, Georgia, U.S.A.

a single exporter among several operating in the Philippines [14]. Two years later, this same supplier exported monkeys to Siena, Italy where another filovirus outbreak subsequently occurred [30]. Sequence analysis of the virus isolated from the monkeys in 1992 showed 0.8% nucleotide changes compared with previous isolates of Ebola-Reston [25]. The mechanism by which EBO-R sustained itself or was re-introduced into the export facility remains unknown. However, within the quarantine facility at Reston, VA, virus was apparently disseminated within and between rooms by the airborne route [6]. Four workers in the facility were also infected, as judged by serological tests and in one case, viral isolation. Fortunately, EBO-R did not cause overt disease in these individuals, unlike the disease in the monkeys which was associated with an 83% fatality rate [14].

While epidemiological date suggest that among humans close contact was usually required for transmission of filoviruses, the possibility of airborne transmission could not always be ruled out [31, 32]. Also, given the clear predilection of EBO-R for the alveoli of naturally infected monkey lung [11], and the demonstrable aerosol infectivity of the human pathogens, Ebola-Zaire (EBO-Z) and Ebola-Sudan (EBO-S) for experimentally exposed monkeys [15–18], investigations of the spread of filovirus infections must consider the possibility of airborne transmission and the possible sources of airborne virus. The studies reported here describe the pathogenesis of EBO-R isolates for experimentally infected cynomolgus monkeys. Analysis of the results gives insight into the pathogenesis of filovirus disease and documents the shedding of infectious virus in various bodily secretions. The disease course of EBO-R in macaques is very similar to human disease following EBO-Z and EBO-S, and this model should facilitate development of effective immunization, treatment, and control strategies.

## Materials and methods

### Viral inocula and assays

Two separate isolates of Ebola-Reston (EBO-R) virus were employed. Both were isolated from monkeys infected during the outbreaks in Reston VA [6, 17]. Monkey #28 virus was associated with the first outbreak (in 1989) and was obtained from a monkey determined to be free of simian hemorrhagic fever (SHF) virus. EBO-R was isolated from the serum of Monkey #28 by inoculation of MA-104 cells as described. Following a second passage in MA-104 cells, this virus was passed once at low multiplicity in vero cells (which are non-permissive for SHF), and then once more in MA-104 cells. AZ-1435 virus was isolated from the spleen of a monkey with signs of respiratory infection [6] during the second outbreak, in March 1990. To reduce the possibility of co-isolating SHF virus, the original 10% spleen homogenate was diluted 1:5 in antiserum with neutralizing antibody against SHF and incubated at 37 °C for 60 min, before further dilution of 1:10 and inoculation of MA-104 cells. The supernatant from this first passaged material was harvested after four days' incubation and passaged at low multiplicity to vero cells, incubated for 11 days. Both virus seeds (Mk #28 and AZ-1435) were demonstrably free of SHF virus as described previously [9]. Infectious virus was assayed by counting plaque forming units (PFU) on monolayers of MA-104 cells maintained under agarose as previously described [16].

Ebola antigen concentrations were measured in sera in a double sandwich capture ELISA using murine monoclonal antibodies to capture Ebola antigens and rabbit antiserum as detector in wells of polyvinyl chloride microtiter plates as described [16].

### Inoculation and treatment of monkeys

Six fully conditioned cynomolgus monkeys (*Macaca fascicularis*), weighing 4.5–6.0 kg were caged individually in stainless steel cages in a maximum containment (BSL 4 laboratory), and fed twice daily with monkey chow supplemented by fresh fruit. Water was available ad libitum. All animals were healthy, and tested seronegative for filovirus reactive antibody. All monkeys were inoculated with 0.5 ml of virus; group II received seem Mk#28, 10 000 PFU sc. Group III received seed AZ-1435, 50 000 PFU sc. To obtain blood, monkeys were lightly sedated with ketamine and bled by femoral venipuncture. Complete necropsies were performed on the five animals that died. For infectivity assays, portions of unfixed tissues were ground with mortar and pestle as 10% w/v suspensions and clarified by centrifugation at $10 000 \times \mathbf{g}$ and stored at $-70\,^{\circ}C$ until assayed for virus.

### Immunological assays

The indirect immunofluorescence antibody test (IFAT) was performed using Ebola virus-infected vero cells dried and fixed onto circular areas of Teflon-coated microscope slides as described [16, 19]. The ELISA assay for IgG responses to EBO-R, Ebola-Zaire (EBO-Z), and Ebola-Sudan (EBO-S) were performed using vero cell lysates as antigens as described [22]. The neutralizing antibody assay was performed by a plaque reduction test, using the constant serum: varying virus format [16]. Neutralizing antibody titers were expressed as a $\log_{10}$ neutralization index (LNI) calculated by the formula $LNI = [\log_{10} (\text{PFU in control}) - \log_{10} (\text{PFU in test serum})]$.

### Necropsy and histology

Complete necropsies were performed on the five animals that died. Representative tissue samples from all organs were selected and immersion fixed in 10% neutral buffered formalin. Tissue specimens were processed and embedded in paraffin. Histology sections were cut at 5–6 μm on a rotary microtome, mounted on glass slides, and stained with Harris's hematoxylin and eosin using a Stainomatic specimen stainer. For immunohistochemical staining, the paraffin-embedded tissues were sectioned at 5 μm and mounted on silane coated glass slides. Following deparaffinization and hydration, the sections were digested with protease and treated with EBO-reactive monoclonal antibodies. Biotinylated horse anti-mouse immunoglobulin immune serum was reacted with the bound monoclonal antibody and the reactive product visualized using a streptavidin and alkaline phosphatase system [17].

### Electron- and immunoelectron microscopy

Tissues from necropsied animals were immersion-fixed in either 2% glutaraldehyde in 0.1 M Millonig's phosphate buffer for standard transmission electron microscopy (TEM) or 2% paraformaldehyde + 0.1% glutaraldehyde for immunoelectron microscopy (IEM). The TEM samples were post-fixed in 1% osmium tetroxide in 0.1 M Millonig's phosphate buffer, rinsed, stained with 0.5% uranyl acetate in ethanol, dehydrated in ethanol and propylene oxide and embedded in POLY/BED 812 resin as described [11, 13]. Ultrathin sections were cut, placed on 200-mesh copper electron microscopy grids, stained with uranyl acetate and lead citrate and examined with a JEOL 100 CX electron microscope (JEOL Ltd., Peabody,

MA) at 80 KV. Samples fixed for IEM were embedded in LR White resin, sectioned, and reacted with murine guinea pig antibodies against Ebola virus, then labeled with anti-guinea pig IgG conjugated with 10 nm gold spheres, as described [13]. Finally, the grids were rinsed in distilled water, stained in uranyl acetate and lead citrate, and examined at 80 KV.

Serum and urine samples were processed for electron microscopy [12] by centrifugation in microcentrifuge tubes at 12 000 × $\mathbf{g}$ for 15 min, removal of the supernatant, resuspension of the virus pellet and transferring to 300-mesh, nickel electron microscopy grids pre-coated with formvar and carbon. Nasal exudates were diluted 1 in 2 with distilled water and applied in 5 µl aliquots to the 300-mesh nickel pre-coated grids. After blotting excess fluid, some grids were processed for IEM as outlined above. Grids for TEM were fixed in drops of 2% glutaraldehyde in PBS for 10 min, exposed to osmium tetroxide vapors, rinsed in drops of distilled water, and negatively stained with 1% phosphotungstic acid (pH = 6.6).

*Hematology, clinical chemistry, and coagulation assays*

Total and differential white blood cell (WBC) counts were obtained using a laser-based (Coulter) hematological analyzer. Leukocyte differentials were confirmed manually using a Wright-stained blood smear. Blood chemistries were measured using a Kodak 700 chemical analyzer. Prothrombin time (PT) and activated partial thromboplastin times (APTT) were performed with an A.C.L. coagulation analyzer. Fibrinogen levels were measured spectrophotometrically using the Clauss method.

## Results

Experimental infection of six cynomolgus monkeys with the filovirus isolated from sick monkeys during the course of the epizootic in Reston, VA, resulted in five deaths and one sublethal infection. The clinical features of the disease resembled the epizootic disease course, and included abrupt anorexia, weight loss, splenomegaly, and fever (> 39 °C). Clinical signs occurred 4 to 5 days after inoculation, followed by death 3 to 9 days later (8 to 14 days after inoculation).

Viremias were detected in all 6 monkeys (Fig. 1). Viremias $> 2 \log_{10}$ PFU/ml were detected by day 3, for 4 of the 5 lethally infected monkeys, and for all by days 5 or 6. Viremia for the one surviving monkey (II-90) was transient, detected only on day 6. In contrast, viremias for the dying monkeys increased to titers $> 7 \log_{10}$ PFU/ml by days 6 to 8, before subsiding prior to death. Antigen was detected by ELISA in serum dilutions of 1:4 or greater when infectious viremias exceeded 2 to $3 \log_{10}$ PFU/ml. Antigenemia curves closely followed infectious viremias until the terminal stages, when antigen titers were maintained at peak values while infectious viremias declined, perhaps as a result of early humoral antibody response. Antibody titers $> 20$, measured as IFAT titers to Ebola-Reston (EBO-R) antigen, were detected as early as day 6 in experiment II, and day 10 in experiment III (Fig. 1). None of the monkeys developed an antibody response to simian hemorrhagic fever (SHF) virus, further evidence that the virus inoculae tested in this experiment were free of SHF virus.

Tissues for 4 of the 5 lethally infected monkeys were tested for infectious virus (Table 1). In the 2 monkeys dying on day 11 (II-95 and III-57), the liver and spleen were clearly sites of viral replication, as were kidney, adrenal gland, and lung. Only one brain was processed for infectivity, and it harbored viral titers no

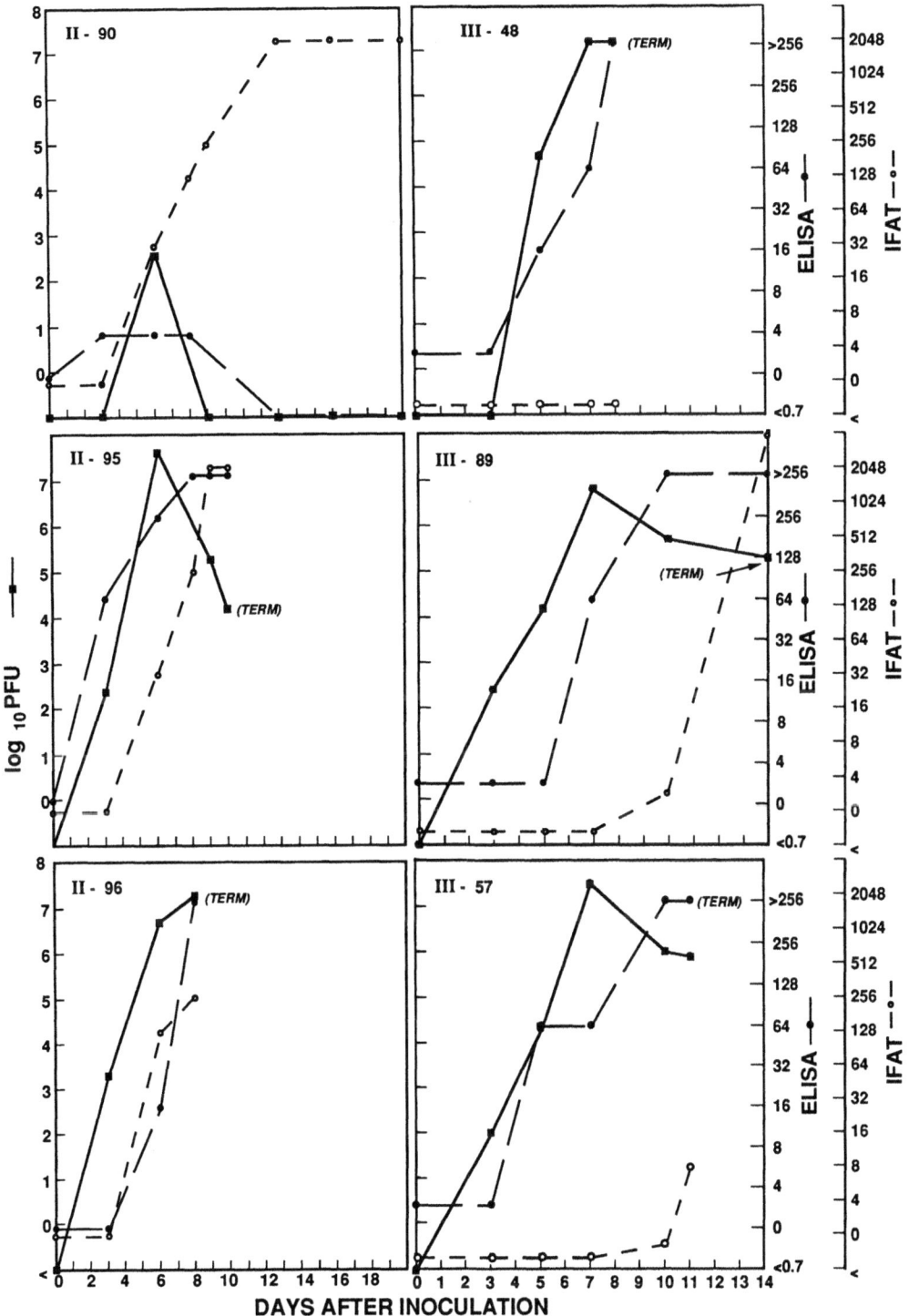

**Fig. 1.** Concentrations of infectious virus, Ebola antigenemia, and IFAT antibody to EBO-R in sera of six cynomolgus monkeys inoculated sc with either of two EBO-R isolates

P. B. Jahrling et al.

**Table 1.** Infectivity of organ homogenates (10% w/v) obtained from terminally
ill monkeys inoculated with Ebola-Reston virus

| Monkey # | Days | $Log_{10}$ PFU/ml or g | | | | | | |
|---|---|---|---|---|---|---|---|---|
| | | serum | liver | spleen | kidney | adrenal | lung | brain |
| III-48 | 8 | 7.2 | 7.1 | 7.5 | 6.2 | 7.0 | 6.4 | ND[a] |
| III-57 | 11 | 5.9 | 7.5 | 7.5 | 7.1 | 7.5 | 6.5 | ND |
| II-96 | 8 | 7.3 | 6.3 | 6.2 | ND | 6.4 | ND | ND |
| II-95 | 11 | 4.2 | 7.0 | 7.1 | ND | 8.2 | ND | 4.2 |

[a] *ND* Not done

higher than the contained blood. Likewise, tissues from the two animals dying
earlier (II-96 and III-48) contained only modest concentrations of infectious
virus relative to blood. However, when tissues from III-48 and III-57 were
examined by transmission electron microscopy (TEM), viral inclusions and
budding viral particles were readily demonstrable in all tissues examined, usually
in association with fixed tissues macrophages and fibroblasts. For example,
virions were visualized in the lung of monkey III-57, in association with the

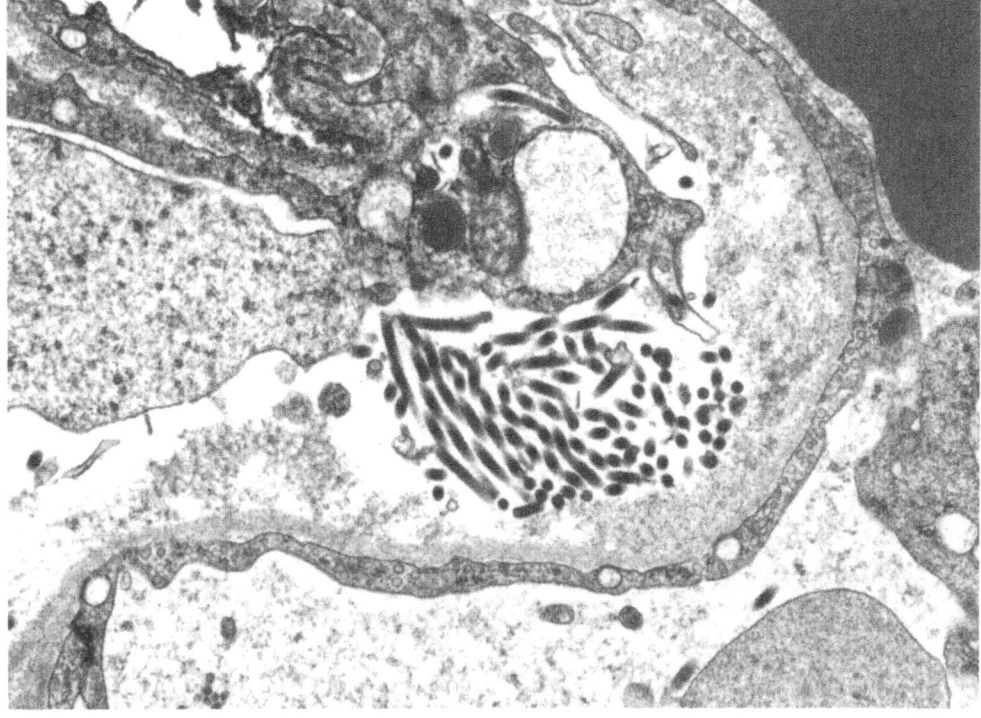

**Fig. 2.** Lung. Virions are present in the interstitium of alveolar septa. Poly/Bed 812 thin
section stained with uranyl acetate and lead citrate. × 29 950

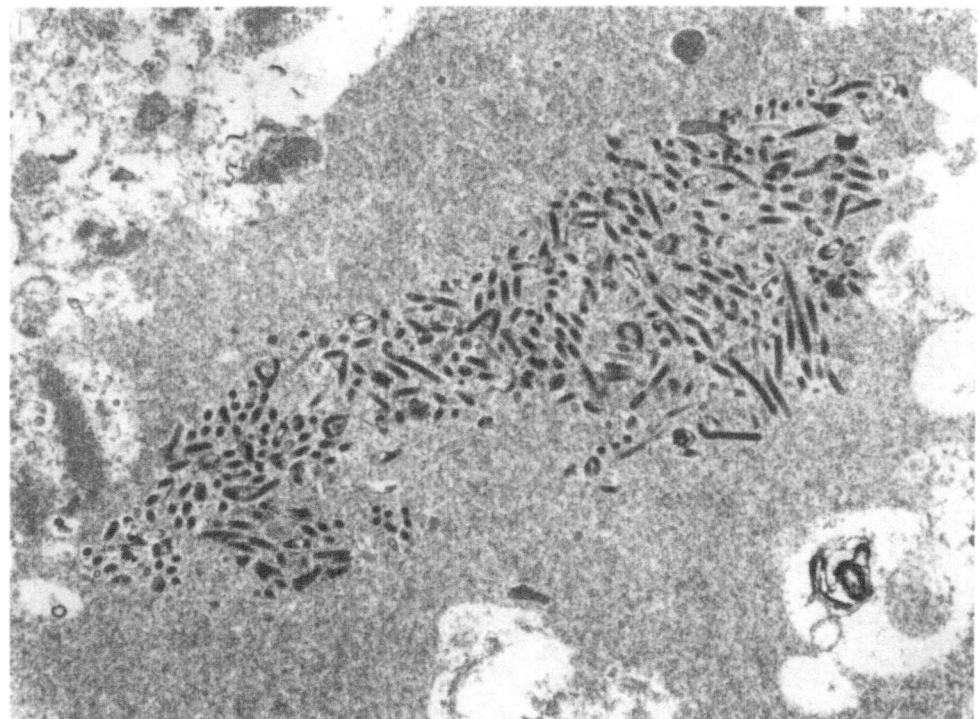

**Fig. 3.** Lung. Thin section through alveoli shows free virions in the pleural fluids. Poly/Bed
812 thin section stained with uranyl acetate and lead citrate. × 17 500

interstitium of the alveolar septae (Fig. 2), and free in the pleural fluid (Fig. 3). In
spleen, virus was found in association with red pulp, but not white pulp. Virions
were interspersed with large deposits of fibrin (Fig. 4). In livers, virus was closely
associated with fibrin deposits in the space of Disse (Fig. 5). Using immunoelectron microscopy (IEM) techniques, virions in the space of Disse were labeled
with gold particles using guinea pig antibodies against EBO-R viral antigens,
(Fig. 6). Mesenteric lymph nodes were congested with large numbers of virions,
although lymphocytes seemed to be uninvolved (Fig. 7).

To determine whether infected animals shed virus, swabs were obtained
sequentially from nasal, pharyngeal, conjunctival, and anal mucosa of animals
III-48, III-57, and III-89 and tested for viral infectivity (Table 2). Pharyngeal and
nasal washes from all three monkeys yielded infectious virus prior to death.
Swabs from conjunctival and anal mucosa of only one animal, III-57, yielded
virus. Direct visualization of viral particles by EM in serum was easily achieved
for all 3 animals, within 5 days of inoculation (Fig. 8). In contrast, viral particles
were detectable by EM in pharyngeal washes and urines only in samples
collected from moribund animals (Fig. 9).

Localization of viral antigens was established immunohistochemically in
mucosal cells of the oral cavity, gastrointestinal tract, and urothelium (Figs. 10, 11).
Electron micrographs confirmed that virus was actively replicating in salivary

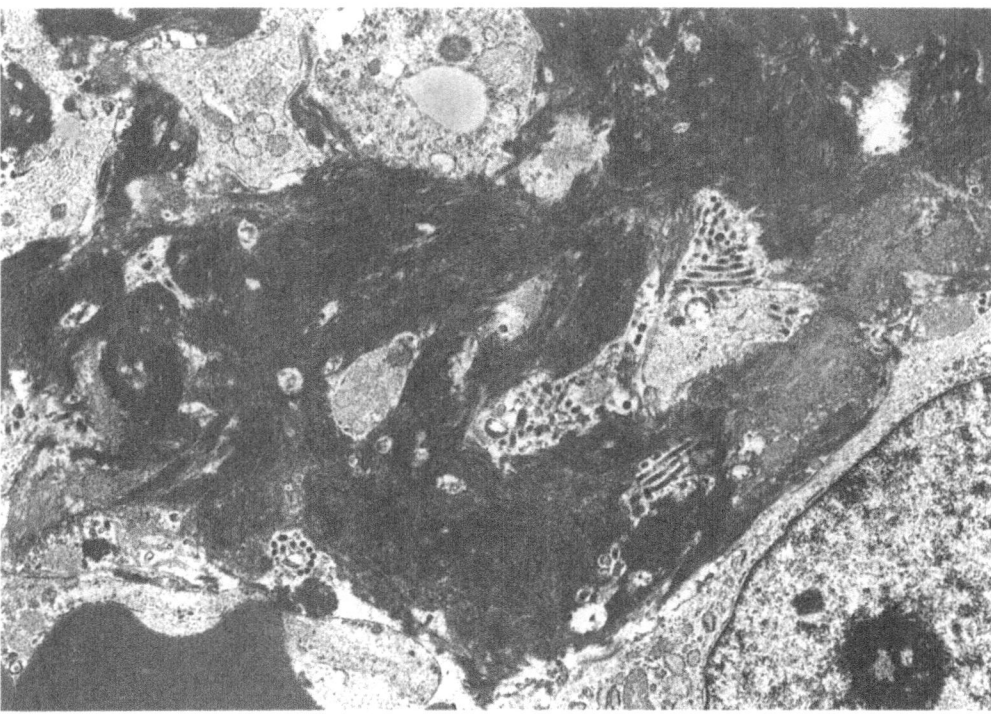

**Fig. 4.** Spleen. This section through red pulp shows several clusters of virions interspersed with large deposits of fibrin. POLY/BED 812 thin section stained with uranyl acetate and lead citrate. × 11 500

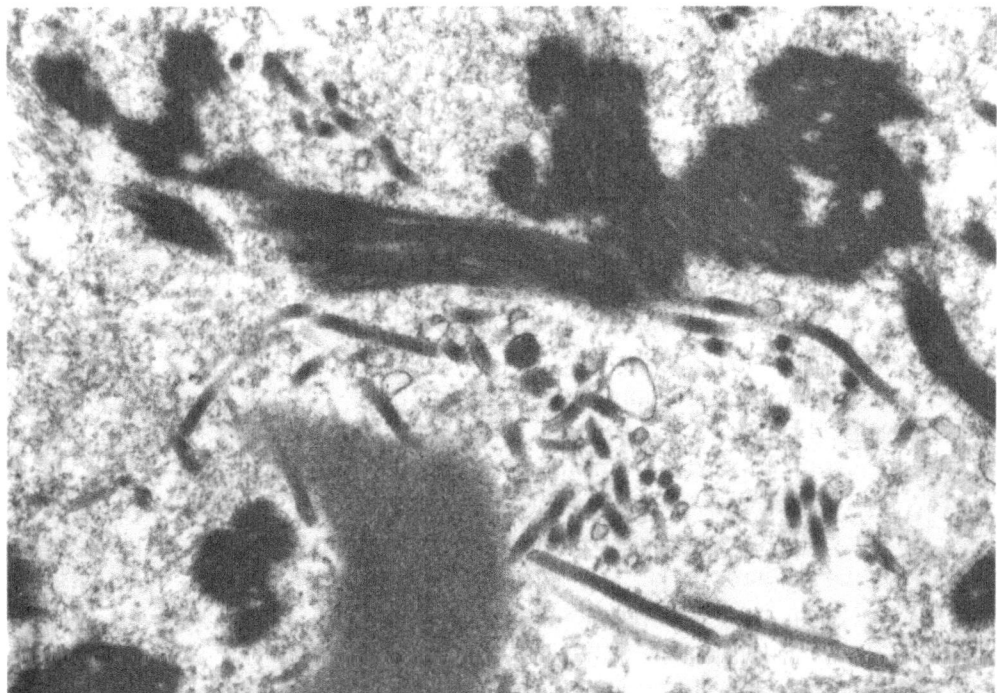

**Fig. 5.** Liver. Virions are closely associated with deposits of fibrin in space of Disse. POLY/BED 812 thin section stained with uranyl acetate and lead citrate. × 31 000

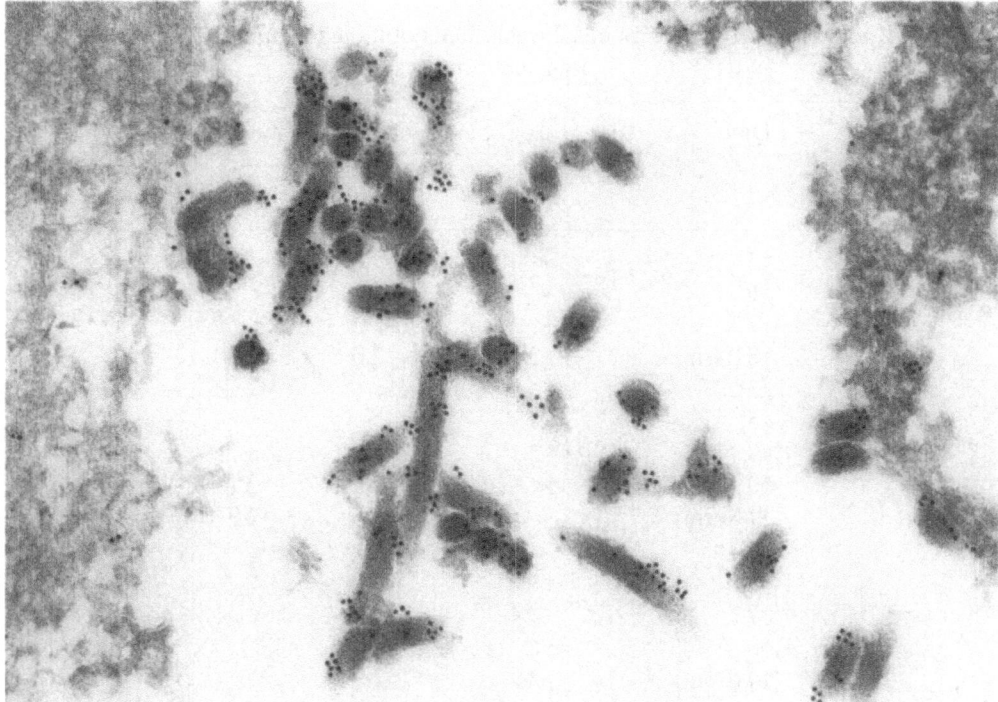

**Fig. 6.** Liver. Positive immunogold staining of virions in the space of Disse of monkey liver when incubated with EBO-R antiserum. LR White thin section stained with uranyl acetate and lead citrate. × 63 250

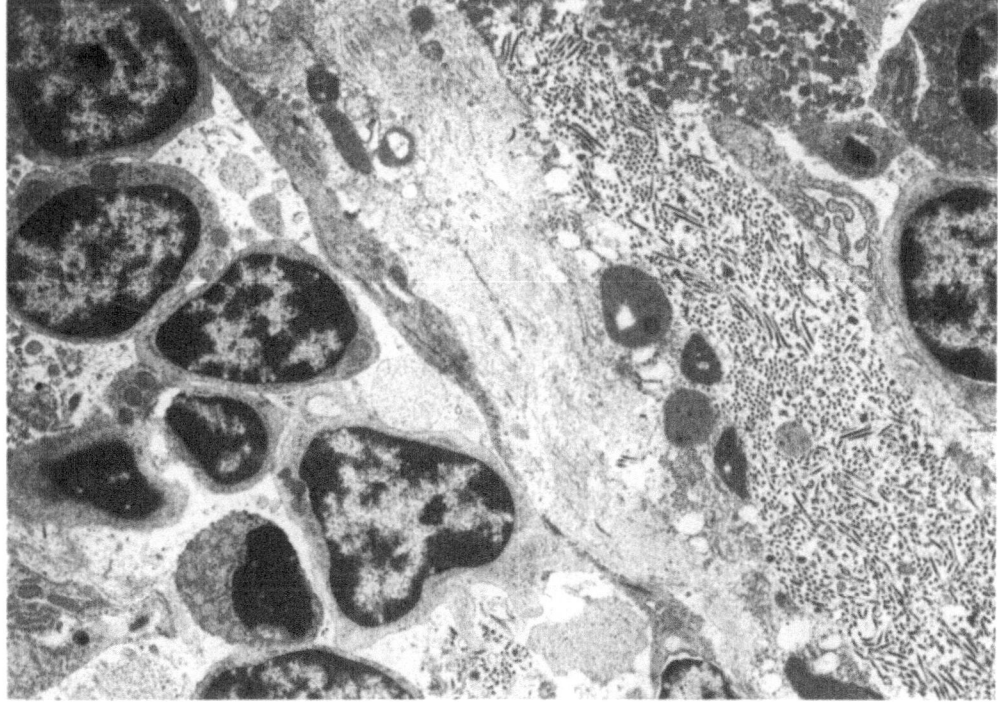

**Fig. 7.** Mesenteric lymph node. Thin section TEM shows medullary sinus congested with a large mass of EBO-R virus and cell debris. Note that lymphocytes show no indication of infection. Poly/Bed 812 section stained with uranyl acetate and lead citrate. × 7 900

**Table 2.** Infectivity of mucosal wash fluids obtained from experimentally
infected monkeys

| Monkey # | Day | $\text{Log}_{10}$ PFU/ml of wash fluid | | | |
|---|---|---|---|---|---|
| | | pharynx | nares | conjunctiva | anal |
| 48 | 3 | <[a] | < | < | < |
| | 5 | < | < | < | < |
| | 7 | < | < | < | < |
| | 8 (term) | 2.3 | 2.0 | < | < |
| 57 | 3 | < | < | < | < |
| | 5 | < | < | < | < |
| | 7 | 2.1 | < | < | < |
| | 10 | < | 2.8 | 2.1 | 2.1 |
| | 11 (term) | < | 3.5 | 2.9 | 2.3 |
| 89 | 3 | < | < | < | < |
| | 5 | < | < | < | < |
| | 7 | 2.3 | < | < | < |
| | 10 | < | < | < | < |
| | 14 (term) | 2.9 | 2.9 | < | < |

[a] < Less than 0.7 $\text{Log}_{10}$ PFU/ml

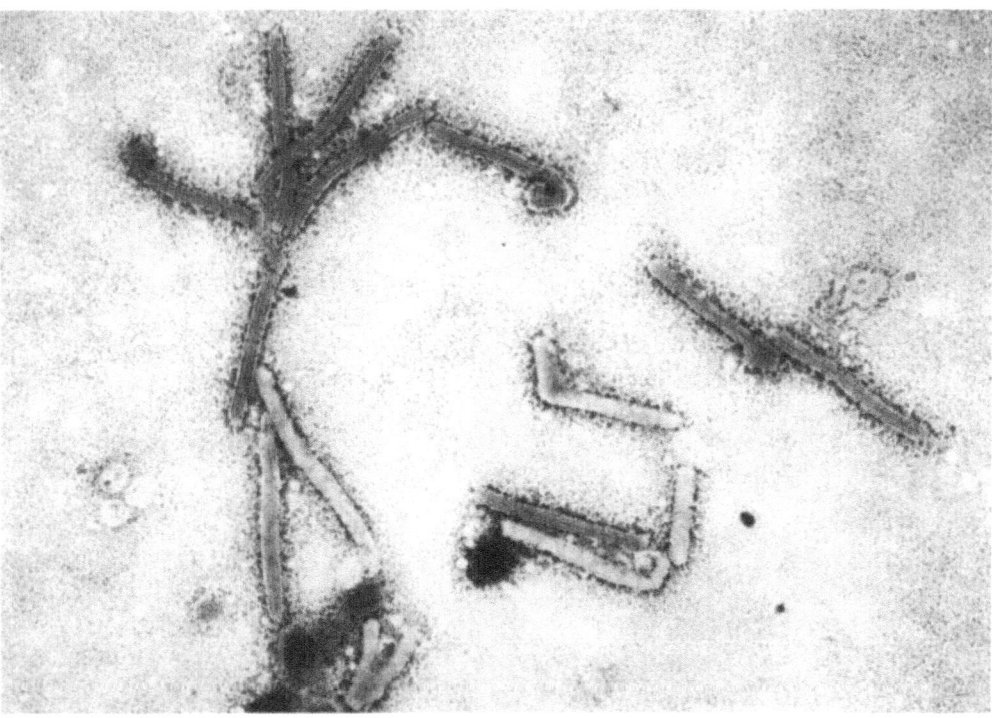

**Fig. 8.** EBO-R virions recovered from serum of infected monkey seven days after inoculation. Preparation negatively-contrasted with 1% phosphotungstic acid (pH = 6.6). × 31 000

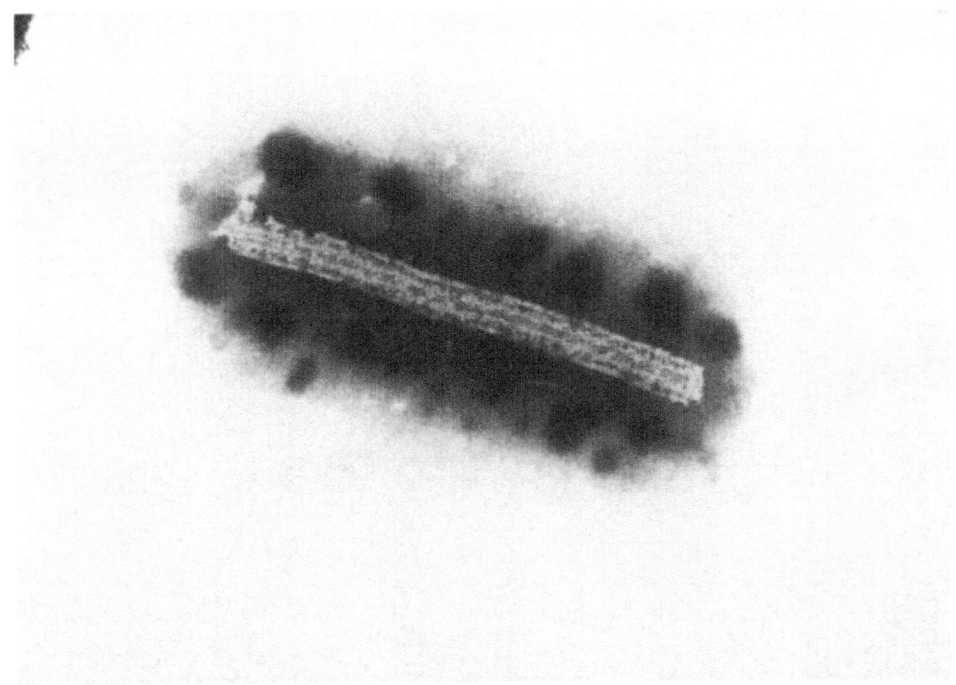

**Fig. 9.** EBO-R virion recovered from nasal exudate of moribund monkey. Preparation negatively-contrasted with 1% phosphotungstic acid (pH = 6.6). × 82 150

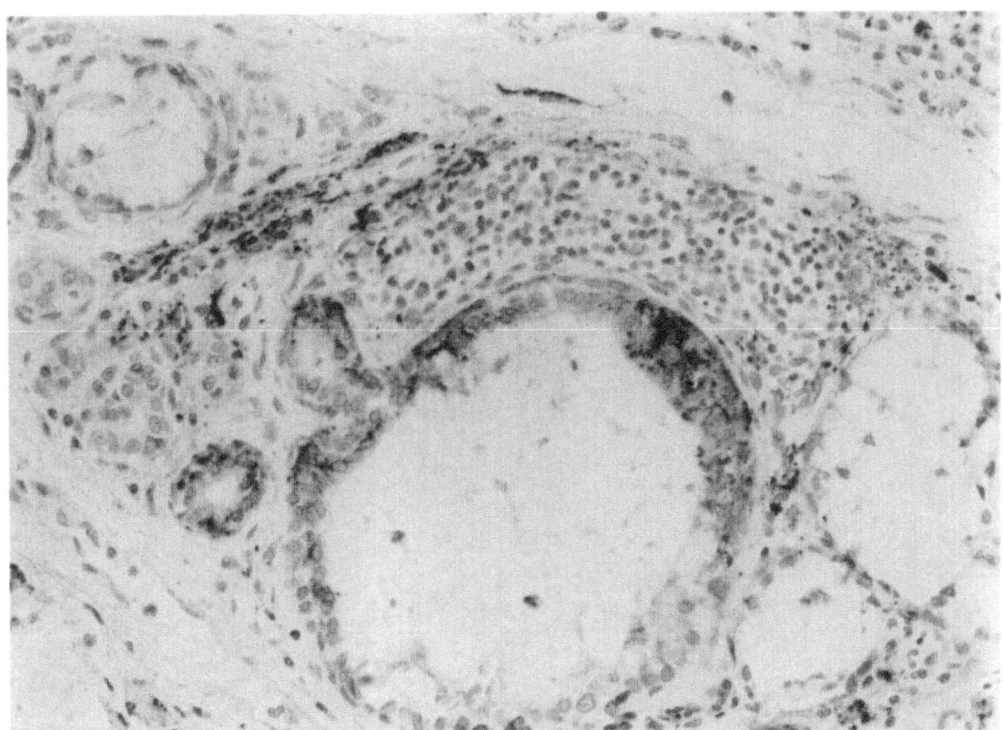

**Fig. 10.** Salivary gland. Immunohistochemical localization of EBO-R antigens in the salivary gland duct epithelium. EBO-R antigen is also present in circulating cells within capillaries and small venules. × 50

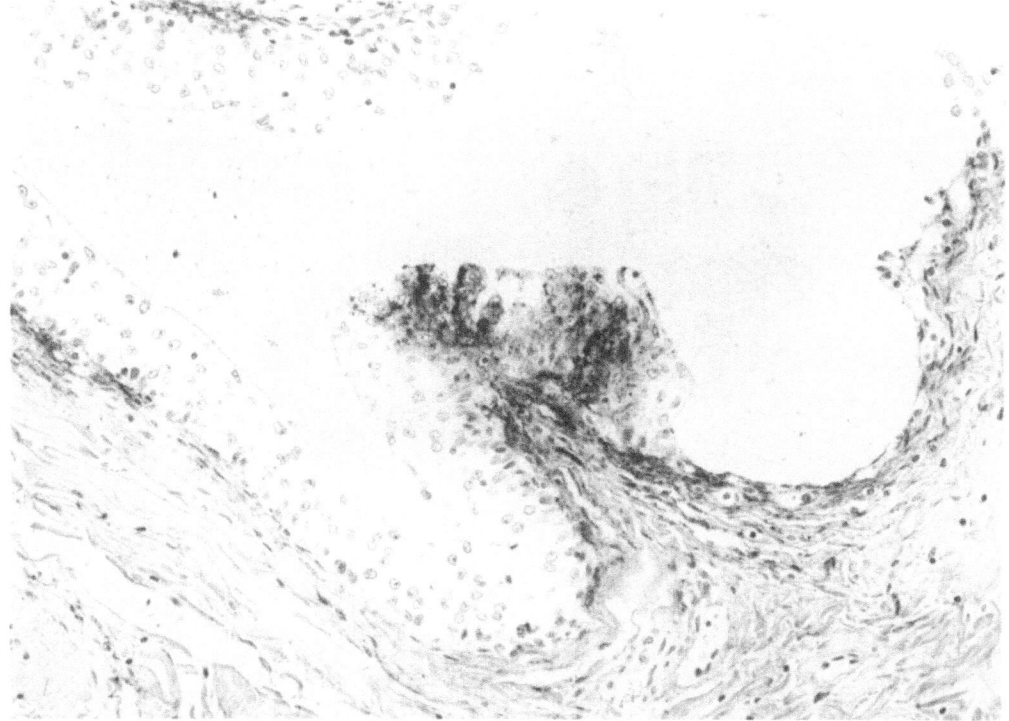

**Fig. 11.** Urinary bladder. Immunohistochemical localization of EBO-R antigens in urothelium. × 50

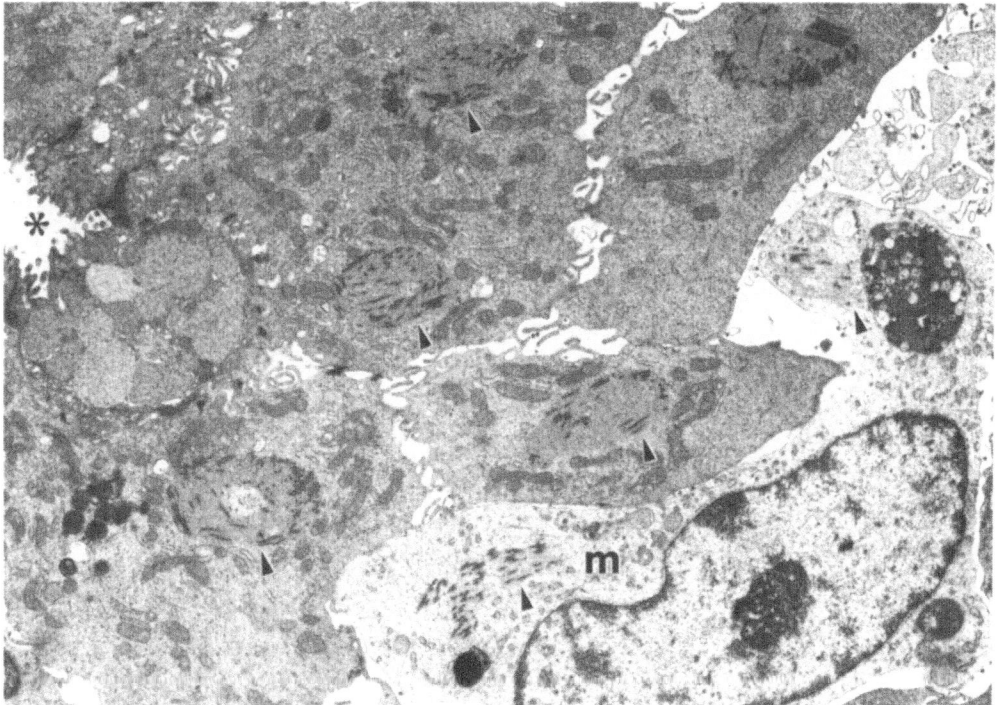

**Fig. 12.** Salivary gland. TEM of sublinguinal salivary gland with EBO-R viral inclusion masses (arrowheads) in epithelial cells lining a duct (*), epithelial cells removed from the duct, and an infiltrating macrophage (m). Poly/Bed 812 thin section stained with uranyl acetate and lead citrate. × 7 380

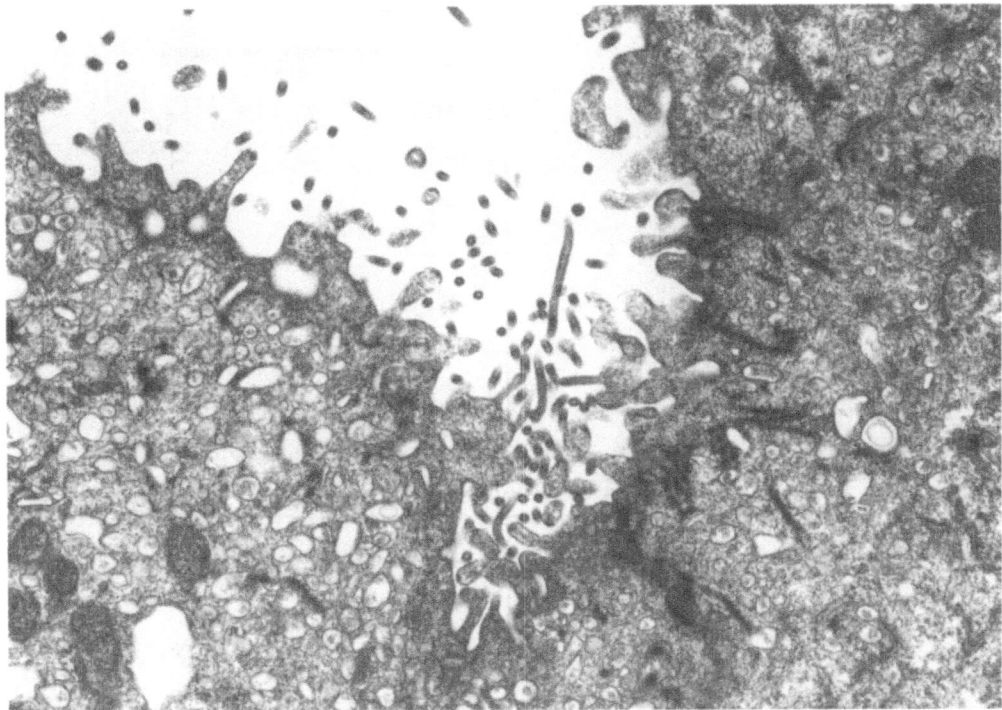

**Fig. 13.** Urinary bladder. Viral-infected epithelial cells lining lumen of urinary bladder, with cytoplasmic viral inclusion material, free virions in the lumen, and viral envelopes protruding into the lumen from plasma membranes. × 25 200

gland (Fig. 12), and urothelium in the bladder (Fig. 13). Thus, it is not necessary to speculate that the source of this infectious virus was infected blood.

Selected coagulation parameters were measured in three of the infected monkeys. Fibrin degradation product (FDP) values increased from a baseline of 0, to 101–202 µg/ml on days 6 to 10. For one monkey (#III-57), values subsequently declined, indicating sytemic clearance of the FDP's. Baseline fibrinogen values (average 185 mg/dl) increased two-fold by day 7 (average 473 mg/dl) (Fig. 14), before declining to values less than baseline at death. Prothrombin times were prolonged from an average of 9.3 sec to 15.3 sec in these animals. Similarly, activated partial thromboplastin times increased from an average of 34.1 sec to more than 60 sec. Platelet counts declined progressively in all animals, and fell below $1.0 \times 10^5$ cells/dl in one animal.

Alterations in several serum enzyme concentrations were pronounced (Fig. 15). The most striking elevations in clinical chemistries occurred in serum lactic dehydrogenase (LDH), serum aspartate amino transferase (AST), alkaline phosphatase (ALP), and creatinine phosphokinase (CPK). Baseline values of LDH increased more than 8-fold in all five animals that died; peak values, which occurred near death in two animals, exceeded 45 000 and 67 000 IU/l Likewise values increased dramatically in the five lethally infected animals, with peak values that exceeded 2 000 IU/l in two animals. Although alkaline aminotrans-

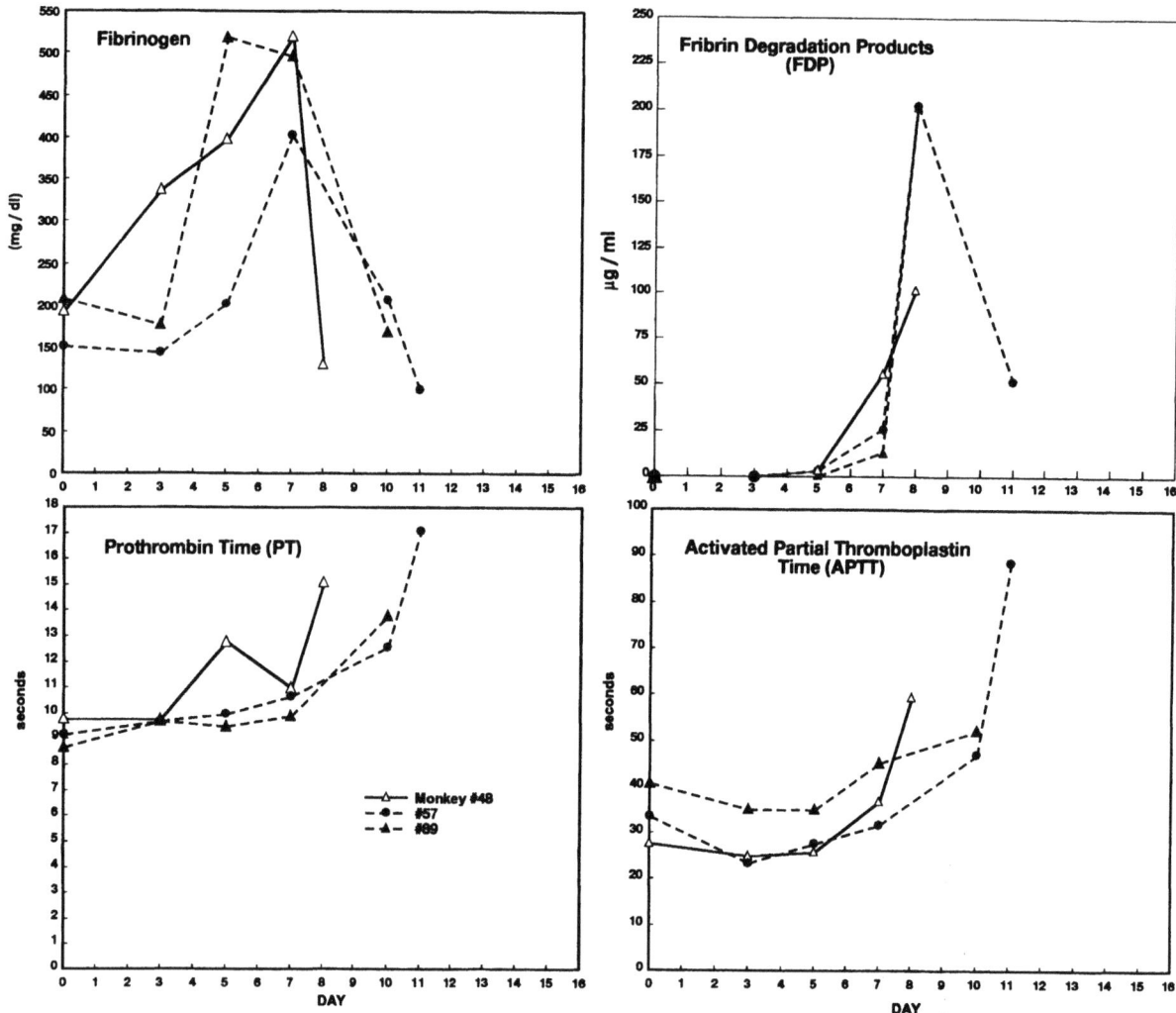

**Fig. 14.** Coagulation parameters measured in plasmas of three experimentally infected cynomolgus monkeys inoculated with EBO-R strain AZ 1435, 50 000 PFU, sc

ferase (ALT) reached an average level of 363 IU/l in terminal animals (range 74–833 IU/l), the ALT levels never exceeded the AST levels at any time. Blood urea nitrogen levels (BUN) remained within normal limits until animals became moribund, at which point they rose dramatically in 3 of the 5 animals, suggesting renal failure occurred as an agonal event.

While all the monkeys developed antibody responses to EBO-R while they were still viremic (Fig. 1), only one monkey (II-90) survived long enough to develop a neutralizing antibody response (Table 3). The earliest detectable response was by IFAT, while the ELISA was somewhat delayed. By both tests, the early response was very specific for EBO-R, with titers 8 to 12-fold higher than for EBO-S and -Z; however, by 9 months, the responses had broadened, and titers varied only 2 to 4-fold. The neutralizing antibody response, measured as the LNI in a virus dilution:constant serum format, was much delayed. The LNI

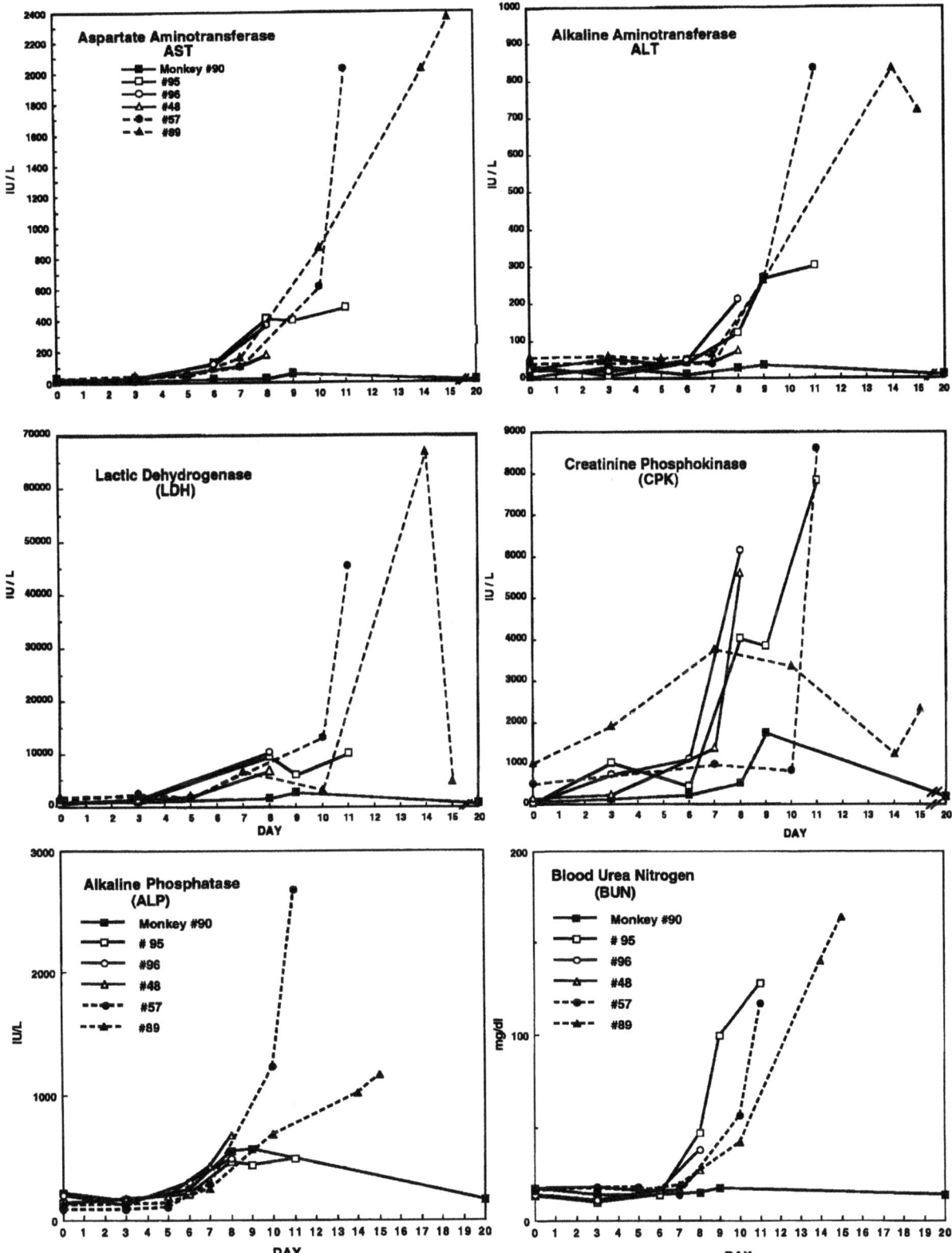

**Fig. 15.** Enzyme concentrations in sera of six cynomolgus monkeys inoculated sc with either
of two EBO-R isolates

**Table 3.** Sequential serology for one monkey (II-90) surviving
Reston viral infection

| Serologic test | Days | Antibody reactivity R[a] | S | Z |
|---|---|---|---|---|
| IFAT[b] | 0 | < 10 | < 10 | < 10 |
| | 8 | 80 | < 10 | < 10 |
| | 13 | 2560 | 160 | 160 |
| | 20 | 2560 | 160 | 320 |
| | 74 | 1280 | 160 | 640 |
| | 181 | 640 | 80 | 40 |
| | 323 | 320 | 160 | 80 |
| | 423 | 320 | 160 | 80 |
| LNI[c] | 0 | < 0.3 | < 0.3 | < 0.3 |
| | 13 | < 0.3 | < 0.3 | < 0.3 |
| | 20 | < 0.3 | < 0.3 | < 0.3 |
| | 30 | 0.5 | < 0.3 | < 0.3 |
| | 74 | 2.3 | 0.4 | < 0.3 |
| | 181 | 3.5 | 0.4 | < 0.3 |
| | 232 | 4.1 | 0.3 | < 0.3 |
| | 323 | > 4.1 | 0.3 | < 0.3 |
| | 423 | > 4.1 | 0.3 | < 0.3 |
| | 423 | > 4.1 | 0.3 | < 0.3 |
| ELISA[d] | 0 | 0.01 | 0.02 | 0.00 |
| (1:100) | 13 | 1.93 | 0.15 | 0.08 |
| | 20 | 1.89 | 0.08 | 0.03 |
| | 323 | 1.93 | 1.79 | 1.19 |
| ELISA | 0 | 0.02 | 0.07 | 0.09 |
| (1:400) | 13 | 1.54 | 0.02 | 0.00 |
| | 20 | 1.05 | 0.05 | 0.01 |
| | 323 | 1.23 | 0.36 | 0.33 |

[a] R Ebola-Reston; S Ebola-Sudan; Z Ebola-Zaire
[b] Numbers represent highest dilution of serum yielding positive immunofluorescence
[c] LNI Log neutralization index, in a constant serum: virus dilution plaque reduction test
[d] Numbers represent optical densities in reaction vessel with serum diluted either 1:100 or 1:400

was first detected on day 30, then gradually increased to reach maximal titers by day 232. Maximal LNI titers were maintained for more than 423 days. Unlike the IFAT and ELISA responses, the LNI did not lose specificity. EBO-R was clearly a distinct serologic entity by cross-neutralization criteria.

## Discussion

The clinical presentation and course of disease in the experimental infection of cynomolgus macaques with EBO-R was very similar to EBO-Z in rhesus and EBO-Z or EBO-S in cynomolgus macaques, although the course of the disease was somewhat longer [1, 9, 10]. All animals experimentally inoculated with the EBO-R isolates were febrile ($>39.5\,°C$) within 4 days, and five of the six inoculated animals died within 8 to 14 days after inoculation. In comparison, monkeys inoculated with EBO-Z and adapted strains of EBO-S frequently died more rapidly, consistently within 8 days of inoculation. Viremias for the lethally infected EBO-R monkeys exceeded $7 \log_{10}$ PFU/ml by day 7, a slight delay in comparison with EBO-Z and EBO-S. EBO-R viremias as well as mortality were higher in this experiment than were reported previously [8, 9]. This discrepancy may reflect the different passage levels of the virus seeds in cell cultures and the higher viral doses used in the present study. Cynomolgus monkeys infected with EBO-R in this experiment developed anorexia, occasional nasal discharge, and splenomegaly, the same findings reported in the cynomolgus monkeys infected naturally with the EBO-R viruses that served as the source for the virus inocula used here [6]. The EBO-R inocula were demonstrably free of SHF virus, whose simultaneous presence during the Reston outbreak complicated epidemiological analyses. EBO-R clearly had the potential to initiate and sustain the monkey epizootic. The true role of SHF virus in the event may never be understood.

Experimentally infected EBO-R monkeys also showed petechial facial hemorrhages and severe subcutaneous hemorrhages in venipuncture sites, analogous to common findings with human Ebola fever [2, 31, 32] which include widespread petechial hemorrhage on the face, chest, and medial aspects of the arms and thighs. EBO-Z also produces these lesions in experimentally infected cynomolgus [9] and rhesus monkeys [1]. EBO-S does so apparently less often [9].

In comparison with other reference filovirus strain infections, the IFAT antibody response to EBO-R occurred much earlier, between 6 to 10 days after inoculation. This was probably not a protective antibody response, however, since all six animals responded but five of them died. Neutralizing antibodies were not detectable until 30 days after infection in the one survivor. The LNI was specific for EBO-R; this antiserum barely neutralized EBO-Z and -S at all, suggesting that EBO-R is a distinct filovirus by serological criteria. The late evolution and persistence of this specific LNI response may have some practical importance. If neutralizing antibody is an important mediator in clearance of viremia and can be used to effectively treat acutely ill patients, efforts should be made to identify plasma donors whose acute illness was several months or more previously, rather than recently convalescent patients. Also, it may well be important to match the subtype of the virus with immune plasma for therapy. EBO-R immune plasma would not be predicted to protect against EBO-Z or -S, and on the basis of preliminary testing in guinea pigs, inappropriately matched antiserum might even precipitate immunologically enhanced disease (P. Jahrling, unpubl. obs.).

In the experimentally infected EBO-R monkeys, the most striking elevations in clinical chemistries occurred with serum lactic dehydrogenase (LDH), serum aspartate amino transferase (AST), alkaline phosphatase (ALP), and cretinine phosphokinase (CPK). These serum enzymes are non-specific markers of cellular injury, and when elevated as a group reflect tissue necrosis in multiple tissues, most notably striated muscle, gastrointestinal tract, lymphoid tissue, and liver. While the ALT values (considered specific markers for hepatocellular damage) were elevated, the ratios of AST:ALT values exceeded 5, suggesting that there are critical extrahepatic viral targets. BUN and creatinine values were markedly elevated, but only in terminal animals, suggesting acute renal failure as a contributing cause of death, and perhaps as a consequence of disseminated intravascular coagulation (DIC).

The coagulation values, presence of FDP's, and declining platelet counts of infected monkeys in this experiment were consistent with a clinical diagnosis of DIC. This diagnosis was confirmed histologically by the presence of fibrin thrombin in tissues of necropsied monkeys, and was reported as a consistent finding in the previous references. While the coagulopathy in EBO-Z infected rhesus monkeys was characterized previously as related to neutrophilia, failure of platelet function and defects in the intrinsic system [10], the exact nature and pathogenesis of the coagulopathy in EBO-R infection will require further investigation. The clinicopathologic picture of sustained neutrophilia, leakage, enzymes indicating multiple tissue damage, and consumption coagulopathy suggest a complex interactive relationship between the three events. Sustained neutrophilic response is uncharacteristic of most viral infections, and in the case of Ebola viruses, may be a response to a combination of chemotactic factors released from tissue damage directly induced by virus infection, ischemic tissue damage occuring secondary to intravascular thrombus formation, and soluble mediators released when platelet aggregation occurs.

Infectious viremia levels exceeded $7 \log_{10}$ PFU/ml of blood, and infectious virus was readily isolated from swabs of the nasal, pharyngeal, conjunctival, and anal mucosa. While the possibility could not be excluded that some of the infectious virus isolated from bodily fluids originated from blood, virus clearly replicated in mucosal cells of the oral cavity, GI and urinary tracts. Likewise, by TEM, Ebola virus replication was demonstrable in these cells and in alveoli, suggesting the potential of generating infectious aerosols from these animals. Thus, virus shed in bodily secretions and in aerosols may be significant sources of contagion. It may also be significant that EBO-Z virus is demonstrably infectious for monkeys exposed by the oral or conjunctival routes [33]. To reduce the spread of Ebola virus, taking precautions against all routes of exposure, including the oral, conjunctival, and aerosol routes are appropriate [2, 6].

One of the two EBO-R isolates (AZ-1435) tested in this study was isolated from a monkey with interstitial pneumonia during the apparent airborne transmission phase of the outbreak at Reston [6]. Although the numbers of experimentally infected monkeys was not statistically significant, AZ-1435 infection was associated with interstitial pneumonia in 3 of 3 inoculated monkeys, while the other EBO-R isolate (H25) was not. This suggests that EBO-R may have accquired a pneumonic potential during the course of the outbreak. The four workers with serologic evidence of infection seroconverted during the later phase; for three of these, aerosol was the presumed route of exposure [4, 26]. Had

EBO-R been as virulent for humans as reference strains of EBO-Z and -S, the combination of pneumonic tropism and virulence could have resulted in multiple aerogenic transmissions and a major human outbreak. Phylogenetically, based on nucleotide sequences, EBO-R is as closely related to EBO-Z and -S as these human pathogens are to each other [7, 25, 29]. There is no guarantee that future filovirus isolates will not possess the attributes of both virulence and aerosol potential.

## References

1. Baskerville A, Fisher Hoch SP, Neild GH, Dowsett AB (1985) Ultrastructural pathology of experimental Ebola hemorrhagic fever virus infection. J Pathol 147: 199–209
2. Bennett D, Brown D (1995) Ebola virus. Br Med J 340: 1344–1345
3. Centers for Disease Control and Prevention (1995) Update: outbreak of Ebola viral hemorrhagic fever-Zaire, 1995. MMWR 44: 381–382
4. Centers for Disease Control (1990) Update: filovirus infection among persons with occupational exposure to nonhuman primates. MMWR 39: 266–267, 273
5. Centers for Disease Control (1995) Management of patients with suspected viral hemorrhagic fever. MMWR 44: 475–479
6. Dalgard DW, Hardy RD, Pearson SL, Pucak GJ, Quander RJ, Zack PM, Peters CJ, Jahrling PB (1992) Combined simian hemorrhagic fever and Ebola virus infection in cynomolgus monkeys. Lab Anim Sci 42: 152–157
7. Feldmann H, Klenk HD, Sanchez A (1993) Molecular biology and evolution of filoviruses. In: Kaaden OR, Eichhorn W, Czerny CP (eds) Unconventional agents and unclassified viruses. Recent advances in biology and epidemiology. Springer, Wien New York, pp 81–100 (Arch Virol [Suppl] 7)
8. Fisher-Hoch SP, Perez-Oronoz GI, Jackson EL, Hermann LM, Brown BG (1992) Filovirus clearance in non-human primates. Lancet 340: 451–453
9. Fisher-Hoch SP, Brammer TL, Trappier SG, Hutwagner LC, Farrar BB, Ruo SL, Brown BG, Hermann LM, Perez-Oronoz GI, Goldsmith CS, Hanes MA, McCormick JB (1992) Pathogenic potential of filoviruses: role of geographic origin of primate host and virus strain. J Infect Dis 166: 753–763
10. Fisher-Hoch SP, Platt GS, Neild GH, Southee T, Baskerville A, Raymond RT, Lloyd G, Simpson DIH (1985) Pathophysiology of shock and hemorrhage in a fulminating viral infection (Ebola). J Infect Dis 152: 887–894
11. Geisbert TW, Jahrling PB, Hanes MA, Zack PM (1992) Association of Ebola-related Reston virus particles and antigen with tissues lesions of monkeys imported to the United States. J Comp Pathol 106: 137–152
12. Geisbert TW, Rhoderick JB, Jahrling PB (1991) Rapid identification of Ebola and related filoviruses in fluid specimens by indirect immunoelectron microscopy. J Clin Pathol 44: 521–552
13. Geisbert TW, Jahrling PB (1990) Use of immunoelectron microscopy to show Ebola virus during the 1989 United States epizootic. J Clin Pathol 43: 813–816
14. Hayes CG, Burans JP, Ksiazek TG, Del Rosario RA, Miranda MEG, Manaloto CR, Barrientos AB, Robles CG, Dayrit MM, Peters CJ (1992) Outbreak of fatal illness among captive macaques in the Philippines caused by an Ebola-related filovirus. Am J Trop Med Hyg 46: 664–671
15. Jaax N, Geisbert T, Jahrling P, Steel K, McKee K, Negley K, Johnson E, Peters C (1995) Preliminary report: natural transmission of Ebola virus (Zaire strain) to monkeys in a biocontainment laboratory. Lancet 346: 1669–1671

16. Jahrling PB (1995) Filoviruses and arenaviruses. In: Murray PR, Baron EJ, Pfaller MA, Tenover FC, Yolken RH (eds) Manual of clinical microbiology, 6th ed. ASM Press, Washington, pp 1068–1081

17. Jahrling PB, Geisbert TW, Dalgard DW, Johnson ED, Ksiazek TG, Hall WC, Peters CJ (1990) Preliminary report: isolation of Ebola virus from monkeys imported to the USA. Lancet 335: 502–505

18. Johnson E, Jaax N, York C, White J, Jahrling PB (1995) Lethal experimental infections of rhesus monkeys caused by aerosolized Ebola virus. Int J Exp Pathol 76: 227–236

19. Johnson KM, Elliott LH, Heymann DL (1981) Preparation of polyvalent viral immunofluorescent intracellular antigens and use in human serosurveys. J Clin Microbiol 14: 527–529

20. Kiley MP, Bowen ETW, Eddy GA, Isaacson M, Johnson KM, McCormick JB, Murphy FA, Pattyn SR, Peters D, Prozesky DW, Regnery RL, Simpson DIH, Slenczka W, Sureau P, Van der Groen G, Webb PA, Wulff H (1982) Filoviridae: taxonomic home for Marburg and Ebola viruses? Intervirology 18: 24–32

21. Ksiazek TG, Rollin PE, Jahrling PB, Johnson E, Dalgard DW, Peters CJ (1992) Enzyme immunosorbent assay for Ebola virus antigens in tissues of infected primates. J Clin Microbiol 30: 947–950

22. Ksiazek TG (1991) Laboratory diagnosis of filovirus infections in nonhuman primates. Lab Anim 20: 34–46

23. Martini GA, Siegert R (eds) (1971) Marburg virus disease. Springer, Berlin Heidelberg New York

24. Murphy FA (1978) Pathology of Ebola virus infection. In: Pattyn SR (ed) Ebola virus haemorrhagic fever. Elsevier/North-Holland, Amsterdam, pp 37–42

25. Peters CJ, Sanchez A, Feldmann A, Rollin PE, Nichol S, Ksiazek TG (1994) Filoviruses as emerging pathogens. Semin Virol 5: 147–154

26. Peters CJ, Johnson ED, Jahrling PB, Ksiazek TG, Rollin PE, White J, Hall W, Trotter R, and Jaax N (1993) Filoviruses. In: Morse S (ed) Emerging viruses. Oxford University Press, New York, pp 159–175

27. Rollin PE, Ksiazek TG, Jahrling PB, Haines M, Peters CJ (1991) Detection of Ebola-like viruses by immunofluorescence. Lancet 336: 1591

28. Sanchez A, Ksiazek TG, Rollin PE, Peters CJ, Nichol ST, Khan AS, Mahy BWJ (1995) Reemergence of Ebola virus in Africa. Emerg Infect Dis 1: 96–97

29. Sanchez A, Kiley MP, Holloway BP, Auperin DD (1993) Sequence analysis of the Ebola virus genome: organization, genetic elements, and comparison with the genome of Marburg virus. Virus Res 29: 215–240

30. World Health Organization (1992) Viral haemmorrhagic fever in imported monkeys. Wkly Epidemiol Rep 67: 142–143

31. World Health Organization/International Commission to Sudan (1978) Ebola hemorrhagic fever in Sudan, 1976. Bull World Health Organ 56: 247–270

32. World Health Organization/International Commission to Zaire (1978) Ebola hemorrhagic fever in Zaire, 1976. Bull World Health Organ 56: 271–293

33. Jaax NK, Davis KJ, Geisbert TW, Vogel AP, Jaax GP, Topper M, Jahrling PB (1996) Lethal experimental infection of rhesus monkeys with Ebola-Zaire (Mayinga) virus by the oral and conjunctival route of exposure. Arch Pathol Lab Med 120: 140–155

Authors' address: Dr. P. B. Jahrling, Scientific Advisor, Headquarters, USAMRIID, Ft. Detrick, Frederick, MD 21702-5011, U.S.A.

Arch Virol (1996) [Suppl] 11: 135–140

# Passive immunization of Ebola virus-infected cynomolgus monkeys with immunoglobulin from hyperimmune horses

P. B. Jahrling, J. Geisbert, J. R. Swearengen, G. P. Jaax, T. Lewis, J. W. Huggins, J. J. Schmidt, J. W. LeDuc*, and C. J. Peters**

United States Army Research Institute of Infectious Diseases Fort Detrick, Frederick, Maryland, U.S.A.

**Summary.** A commercially available immunoglobulin G (IgG) from horses, hyperimmunized to Ebola virus, was evaluated for its ability to protect cynomolgus monkeys against disease following i.m. inoculation with 1 000 PFU Ebola virus (Zaire '95 strain). Six monkeys were treated immediately after infection by i.m. injection of 6.0 ml IgG; these animals developed passive ELISA titers of 1:160 to 1:320 to Ebola, two days after inoculation. However, the beneficial effects of IgG treatment were limited to a delay in onset of viremia and clinical signs, in comparison with untreated controls. The six IgG recipients had no detectable viremia day 5, in contrast with three virus infected controls whose viremias exceeded $7.0 \log_{10}$ PFU/ml that day. The controls died on days 6, 6, and 7, while two IgG recipients died day 7 and the remaining 4 died day 8, all with high viremias. These results document that passively acquired antibody can have a beneficial effect in reducing the viral burden in Ebola-infected primates; however, effective treatment of human patients may require antibodies with higher specific activities and more favorable pharmacokinetic properties than the presently available equine IgG.

## Introduction

The recent outbreak of Ebola fever in Zaire claimed 244 human lives (MMWR 95) among 344 cases and attracted world-wide attention to the highly virulent nature of this viral infection [2, 3, 13]. Unlike some other viral haemorrhagic fevers (VHF) for which effective antiviral drugs have been identified, Ebola virus infections can not be effectively treated [11]. Immune plasma has been used successfully with other VHF infections [8, 9, 11] and there are anecdotal case reports in the literature suggesting the potential benefits of passive immunization

Present addresses: * World Health Organization, Geneva, Switzerland; ** Special Pathogens Branch, Centers for Disease Control, Atlanta, Georgia, U.S.A.

against Ebola infection [1]. In the waning days of the 1995 outbreak, reports began to surface that whole blood transfusions from recently convalescent patients had been used with remarkable success to treat acutely ill Ebola fever patients. Evaluation of those studies must await formal publication and analysis of the virological and immunological aspects of the treated patients in comparison with untreated cohorts. Even if these whole blood transfusion studies withstand scientific scrutiny and prove the principle of antibody therapy for Ebola fever, the problems of acquiring, processing, and storing this material would remain [6].

In June 1995, the World Health Organization (W.H.O.) was made aware of the existence of hyperimmune immunoglobulin G (IgG) against Ebola, prepared by Russian investigators in horses [6]. This product had been reported to protect baboons experimentally infected with Ebola virus [10] and was offered by the Russian Association "Epidbiomed" for sale to the W.H.O. to be used in Zairian patients. The W.H.O. obtained several hundred doses of this product, and requested that the United States Army Research Institute of Infectious Diseases (USAMRIID), in its capacity as a W.H.O. collaborating Center-Designee with BL-4 primate capability evaluate this product in cell culture and in animal models. Purified equine IgG would circumvent many of the problems associated with human plasma acquisition, and would probably contain higher concentrations of neutralizing antibodies. This report documents the initial evaluation of the "Epidbiomed" hyperimmune IgG product.

## Materials and methods

The IgG was received in 3 ml glass vials from Epidbiomed, labeled, "Immunoglobulin against Ebola fever, from horse antiserum, liquid, (basic preparation), Lot N34". These vials contained colorless, weakly opalescent liquid. Total IgG was determined by nephthalometry, based on antigen-antibody binding between the sample and anti-horse IgG. Purified horse IgG was the standard. Reagents were from Cappel Laboratories, Inc. The sample was devoid of hemoglobin; total protein was 140 mg/ml, and total IgG was 120 mg/ml. By size exclusion chromatography using Pharmacia Superose-12 and Superose-6 columns and a Pharmacia FPLC instrument, chromatographic profiles and elution profiles of the sample were compared with reference (Cappel) equine IgG. On both columns, the sample gave a single symmetrical peak with an elution volume identical to the IgG standard. No albumin or higher molecular weight materials (aggregates) were found in the sample. Sodium dodecyl-sulfate-polyacrylamide gel electrophoresis (SDS-PAGE) of the IgG sample gave the expected peaks for heavy and light chains; it was free of albumin, and exhibited a high degree of purity.

The anti-Ebola titers of the equine IgG were measured in several standard serological tests. The indirect immunofluorescent antibody titer (IFAT) was measured using Ebola-Zaire infected vero cells [4] and amounted to 1:20 480. By ELISA, using Ebola-Zaire infected vero cell lysate as antigen [7], the IgG titered 1:256 000. The neutralizing antibody titers were determined by plaque reduction in both the serum dilution test (PRNT) against 100 plaque forming units (PFU) of Ebola-Zaire (Mayinga strain) and in the virus dilution test using a constant serum dilution (1:100) which yields a $Log_{10}$ neutralization index (LNI) as described [4]. The 80% PRNT titer was 1:2 560, and the 50% PRNT was 1:10 240. The LNI was 4.2 against both Ebola-Mayinga and against a 1995 isolate of the virus (#807224) obtained from the Centers for Disease Control, Atlanta, GA. These neutralizing antibody titles are much higher than any we have ever measured in human convalescent plasma. Passive equine IgG titers against Ebola virus were determined by a modification of the standard procedure [7] using goat anti-horse IgG. Quantitation of total horse IgG in monkey plasma was by ELISA, using a reference standard (Cappel Laboratories, Durham, NC, USA) to generate a standard curve.

Cynomolgus monkeys were used to measure protective efficacy of the IgG. Nine monkeys, weighing 5–6 kg, were anesthetized with Telazol and inoculated i.m. in the leg with 0.5 ml containing 1 000 pfu of Ebola (Zaire '95) virus. Three of these monkeys served as untreated, virus controls. The remaining six monkeys were treated immediately following virus administration by inoculation of 6.0 ml undiluted equine IgG, as supplied by Epidbiomed, (2.0 ml in each of three sites, both arms plus the leg opposite that inoculated with virus). Monkeys were supplied with food and water ad libidum and housed in individual cages in the BL-4 facility. They were anesthetized with Telazol on days 2, 5, and 7 following virus inoculation, (or, when terminally ill, on days 6 or 8) and bled for determination of infectious circulating virus, passive antibody titers, and standard hematological and clinical pathology parameters. Temperatures were also determined with a rectal probe while the animals were being bled. All animals were closely monitored twice daily and were scored objectively for a number of clinical parameters. All terminally ill monkeys were euthanized and necropsied for pathological examination. Viral infectivity assays on plasma and tissue homogenates were performed as described by plaquing on vero cell monolayers [5].

## Results

The beneficial effects of IgG treatment were limited to a delay in onset of viremia and clinical signs, and a prolongation in mean time to death (from 6.3 days in untreated controls to 7.7 days in IgG recipients). Five days after Ebola virus inoculation, the three cynomolgus monkeys that served as untreated controls were febrile ($> 40\,°C$) and anorectic; all three had developed moderate to severe petechial rashes on the arms and lateral thorax. In contrast, none of the six virus-infected IgG recipients displayed any visible signs of illness on day five. Most impressively, the treated group had no detectable viremia on day 5, in contrast with control animals whose viremia exceeded $7\log_{10}$ PFU/ml that day (Fig. 1a). However, 5 of the 6 treated animals were febrile (range 38.6–40.1 °C) and hematologic data revealed a leukocytosis (24 500 WBC/mm$^3$, 87% segmented neutrophils), like untreated controls. On day 6, two of the three controls died, while all of the IgG recipients became symptomatic. By day 7, the third control had died, and all six treated animals were viremic ($> 7\log_{10}$ PFU/ml) and febrile (range 38.5–40.6 °C). Two IgG recipients died day 7, while the remaining four animals died day 8 (Fig. 1a).

Comparison of increases in clinical chemistry values (including AST, GGT, and BUN) between groups confirmed the impression that IgG treatment served only to delay the onset of disease. Comparison of infectious viral burdens in target tissues of treated versus control animals indicated no significant effect of IgG on quantitative or qualitative distribution of virus, and preliminary histopathologic examination of tissues revealed no suggestion of immunologically mediated disease in IgG recipient animals.

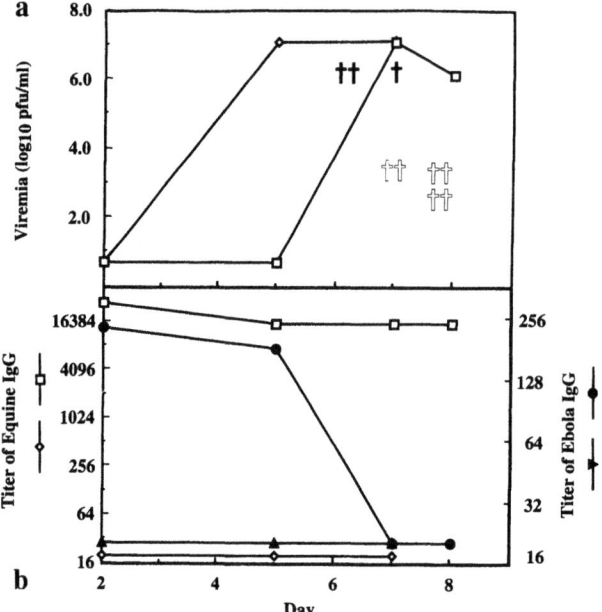

**Fig. 1. a** Geometric mean titer of virus in plasmas from three Ebola virus-infected cynomolgus monkeys (controls, open triangles) and six virus-infected IgG recipients (open squares). Days of death are recorded for controls (black crosses) and IgG recipients (open crosses). **b** Geometric mean antibody titers in virus-infected IgG recipients and controls. Total equine IgG in recipient animals ranged from 1:16 000 to 1:32 000 day 2 (open squares) and was sustained through day of death, day 7 or 8. Total equine IgG was undetectable in controls (open triangles). Specific equine Ebola ELISA titers ranged from 1:160 to 1:320 on day 2 in IgG recipients (black circles); these titers were sustained through day 5 but then declined to undetectable titers day 7. Passive Ebola IgG titers in controls remained undetectable (black triangles)

The IgG recipient animals acquired passive anti-Ebola (equine) IgG titers ranging from 1:80–1:320 on day 2 (geometric mean = 226) (Fig. 1b). The geometric mean titer of total equine IgG was 1:28 508 (or 108 µg/ml) on day 2. The titer of total equine IgG was sustained above 1:16 000 (> 52 µg/ml) through day 8, when the animals died. In contrast, the specific anti-Ebola antibody titers were sustained only through day 5, but then declined to undetectable levels by day 7. This disappearance of specific anti-Ebola antibody coincided with the dramatic evolution in viremia between days 5 and 7 in the IgG recipient animals. None of the animals developed active immune responses to Ebola before they died.

The kinetics of passive IgG disappearance suggested that a second infusion of IgG around day 5 might be beneficial. In uninfected monkeys inoculated with 6 ml of equine IgG, we determined that passive IgG titers (both total and Ebola-specific) were maintained through day 8 at titers of 1:16 000 and 1:160, respectively, then disappeared on day 9, suggesting immunological clearance. Although testing the potential benefit of a second IgG infusion on day 5 has merit, the possibility of inducing serum sickness must be considered. Also, the rapid evolution of viremia between days 5 and 7 suggests that target tissues were already virologically involved on day 5. Humoral antibody is unlikely to be effective in clearing virus from infected tissues. Moreover, the total infectious viral burden in the entire animal is likely to exceed the neutralizing capacity of the IgG infused on day 5. The ability of IgG to prevent infection of target tissues and to prevent spread, once infection is established, may be clarified by immunohistological assessment of viral antigen distribution in target tissues from sequential sacrifice studies, now being considered.

## Discussion

The results of protective efficacy studies must be interpreted with caution. While the present studies demonstrated failure of the equine IgG to protect cynomolgus monkeys against lethal Ebola virus infection, previously published studies using small numbers of Hymadryl baboons suggested the opposite [8]. In those experiments, two baboons receiving the same dose of IgG (6 ml/6 kg baboon) two hours prior to virus inoculation survived; two of three baboons treated with IgG immediately after infection also survived. Cynomolgus monkeys develop higher viremias and die more quickly with higher viral titers in target tissues than baboons infected with Ebola; thus the odds of treatment success may be less in the cynomolgus monkey model. The strain and dose of Ebola virus may also be a factor; the present studies used a recent Ebola virus isolate from the 1995 outbreak in Zaire, and the dose (1 000 PFU, i.m.) could be somewhat higher than the 30 $LD_{50}$ (for African green monkeys) dose used in the baboon study [8]. In preliminary studies using strain 13 guinea pigs infected with a guinea pig-adapted strain of Ebola-Zaire (Mayinga), we demonstrated that animals treated with the equine IgG (1 ml/kg, i.m.) simultaneously with injection of 1 000 $GPLD_{50}$ survived, with no detectable viremia but eventual seroconversion to the infecting virus strain. It was on this basis that we initiated the monkey studies. However, we subsequently learned that when treatment of guinea pigs was delayed until the onset of detectable viremia (day 4), IgG was totally ineffective in preventing death, and even failed to clear the modest (3–4 $log_{10}$ PFU/ml) viremia.

The equine IgG would thus be predicted to have marginal effectiveness in treating human patients with acute Ebola virus infections. In both the published baboon studies [8] and the cynomolgus monkeys and guinea pig studies reported herein, the dose of IgG was 1 ml/kg. This is ten-fold higher than the dose recommended by the manufacturer for human treatment. There is no reason to postulate that lower doses of IgG would be more effective than higher doses in animal models, further eroding confidence in the efficacy of this IgG for treatment of human patients. However, in the absence of any data regarding adverse clinical reactions in human patients, the potential benefit of this IgG in suppressing the infectious viremia might drive a recommendation to use it in an acutely ill patient. Recent verbal reports that patients were effectively treated with whole blood transfusions from recently convalescent patients in Zaire suggest that quantities of antibody, predicted to be marginally effective in laboratory tests, may still be beneficial in actual human cases. If neutralizing antibody is the protective entity, then the equine IgG (with an LNI = 4.1) is clearly superior to the majority of convalescent human plasma (with LNI < 1.0).

Perhaps the principal value of the equine IgG is that for the first time it has been possible to demonstrate that passively acquired antibody has a beneficial effect in reducing the viral burden in an Ebola-infected primate. Thus, investment in strategies using human monoclonal antibodies, which may have higher specific activities and more favorable pharmacokinetic properties than equine IgG, is clearly indicated.

## References

1. Bowen ETW, Lloyd G, Platt G, Mcardell LB, Webb PA, Simpson DIH (1978) Virological studies on a case of Ebola virus infection in man and monkeys. In: Pattyn SR (ed) Ebola virus haemorrhagic fever. Elsevier/North-Holland Biomedical Press, Amsterdam, pp 95–100
2. Centers for Disease Control and Prevention (1995) Outbreak of Ebola viral haemorrhagic fever-Zaire, 1995. MMWR 44: 381–2
3. Centers for Disease Control and Prevention (1995) Update: Outbreak of Ebola viral haemorrhagic fever-Zaire, 1995. MMWR 44: 468–475
4. Jahrling PB (1995) Filoviruses and arenaviruses. In: Murray PR, Baron EJ, Pfaller MA, Tenover FC, Yolken RH (eds) Manual of clinical microbiology, 6th ed. ASM Press, Washington, pp 1068–1081
5. Jahrling PB, Geisbert TW, Dalgard DW, Johnson ED, Ksiazek TG, Hall WC, Peters CJ (1990) Preliminary report: isolation of Ebola virus from monkeys imported to the USA. Lancet 335: 502–505
6. Johnson KM, Webb PA, Heymann HL (1978) Evaluation of the plasmapheresis program in Zaire. In: Pattyn SR (ed) Ebola virus haemorrhagic fever. Elsevier/North-Holland Biomedical Press, Amsterdam, pp 219–222
7. Krasnyanskii VP, Mikhailov VV, Borisevich IV, Gradoboev VN, Evseev AA, Pshenichnov VA (1994) Preparation of hyperimmune horse serum to Ebola virus. (1994) Vopr Virus 2: 91–92
8. Ksiazek TG (1991) Laboratory diagnosis of filovirus infections in nonhuman primates. Lab Anim 20: 34–46
9. Keane E, Gilles HM (1977) Lassa fever in Panguma hospital, Sierra Leone, 1973076. Br Med J 1: 1399–1402
10. Maiztegui JI, Fernandez NJ, deDamilano AJ (1979) Efficacy of immune plasma in treatment of Argentine haemorrhagic fever and association between treatment and a late neurological syndrome. Lancet 2: 1216–1217
11. Mikhailov VV, Borisevich IV, Chernikova NK, Potryvaeva NV, Krasnyanskii VP (1994) An evaluation of the possibility of Ebola fever specific prophylaxis in baboons (Papio hamadryas). Vopr Virus 2: 82–84
12. Monath TP (1990) Ribavirin, Interferon, and antibody approaches to prophylaxis and therapy for viral haemorrhagic fever. Curr Opin Infect Dis 3: 824–833
13. Muyembe T, Kipasa M [for the International Scientific and Technical Committee and WHO Collaborating Centre for Haemorrhagic Fevers] (1995) Ebola haemorrhagic fever in Kikwit, Zaire (Correspondence). Lancet 345: 1348
14. Sanchez A, Ksiazek TG, Rollin PE, Peters CJ, Nichol ST, Khan AS, Mahy BWJ (1995) Reemergence of Ebola virus in Africa. Emerg Infect Dis 1: 96–97

Authors' address: Dr. P. B. Jahrling, Scientific Advisor, Headquarters, USAMRIID, Ft. Detrick, Frederick, MD 21702-5011, U.S.A.

Arch Virol (1996) [Suppl] 11: 141–168

© Springer-Verlag 1996

# Patients infected with high-hazard viruses: scientific basis for infection control

**C. J. Peters[1], P. B. Jahrling[2], and A. S. Khan[1]**

[1] Special Pathogens Branch, Division of Viral and Rickettsial Diseases, National Center for Infectious Diseases, Centers for Disease Control and Prevention, Atlanta, Georgia, [2] United States Army Medical Research Institute for Infectious Diseases, Ft. Detrick, Frederick, Maryland, U.S.A.

**Summary**. Most of the viral hemorrhagic fevers (VHFs) are caused by viruses that are handled in high containment laboratories in Europe and the United States because of their high pathogenicity and their aerosol infectivity. Special precautions should be taken when caring for patients infected with these viruses, but most hospitals can safely provide high-quality care. The major danger is parenteral inoculation of a staff member. Fomites and droplets must be considered as well. The role of small particle aerosols in inter-human transmission continues to be controversial. We believe that the aerosol infectivity observed for these viruses in the laboratory and the rare clinical situations that suggest aerosol spread dictate caution, but the many instances in which no transmission occurs provide a framework in which a measured approach is possible. The major challenge is in early recognition by an educated medical staff and rapid specific etiologic diagnosis.

## Introduction

In the last year, there has been increasing recognition by the general medical community and the public of two basic facts: many viral diseases exist worldwide and modern rapid travel can transport these diseases anywhere on the globe within their incubation period. The introduction of exotic viruses into human populations is usually accompanied by several concerns, including the risk of inter-human transmission and the possibility of the viruses establishing themselves in the new environment. There is also an almost inevitable fear factor compounded by a lack of knowledge about the disease by medical science in general and particularly at its site of distant introduction. We will attempt to give some perspective to these concerns and to develop principles that can assist in the management of such virus infections. The viruses we will discuss are those that cause a syndrome referred to as viral hemorrhagic fever (VHF). This is a severe multi-system disease with diffuse vascular damage and dysregulation often accompanied by hemorrhage; the bleeding manifestations are generally a reflection of widespread tissue involvement rather than being life-threatening in volume. Most of the viruses that cause this syndrome are aerosol infectious and therefore pose a particular hazard in the laboratory and perhaps in the medical care setting.

The viruses commonly associated with the VHF syndrome are listed in Table 1. They are all zoonotic, lipid-enveloped RNA viruses. Several viruses will not be considered further. For example, dengue virus is not a highly virulent organism and is not associated with inter-human transmission; therefore, the only precaution needed for dengue patients is screening from mosquitoes to prevent further arthropod transmission. Viruses of the genus *Hantavirus*, such as Hantaan virus (causing HF with renal syndrome) and Sin Nombre virus (causing hantavirus pulmonary syndrome), are hazardous in the laboratory and particularly so when working with inoculated rodent reservoirs. However, patients infected with these agents have never been associated with inter-human transmission, presumably because the immune response occurs at the onset of disease and reduces viral titers ([44], C. Vitek, unpubl. data), and no special precautions are needed with these patients.

The remaining diseases provide a difficult problem to evaluate. The incidence of person-to-person transmission varies greatly, and nosocomial transmission has ranged from common and predictable to rare and unpredictable. The potential transmission mechanisms include direct contact or direct projection of droplets onto mucous membranes; indirect transmission via fomites and body fluids; and airborne transmission via small-particle aerosols. The viruses that cause these diseases are all infectious by aerosol in the laboratory, and because of their lethal potential for humans require either biosafety level (BSL)-4 containment or BSL-3 containment and vaccination (Table 1) [72]. Nonetheless, the absolute risk to family members and medical staff is low and it is by no means established that aerosols are a major consideration in inter-human transmission.

To approach the problem of when and why such patients pose a health risk to those around them, it is necessary to evaluate the sparse field data available from human infections, consider information gleaned from pathogenesis experiments in animals, review little-appreciated facts surrounding small-particle aerosols, and examine previously published guidelines [11, 16]. This information can be synthesized into an approach to the care of infected patients who may present far from the area where they acquired infection.

### Human data

The ultimate test of the risk of caring for patients with VHF comes from direct observations in hospitals, but these data are scarce, derived under varying conditions of medical management, and often ambiguous. Thus, we also take into account surrogates in our decision process. These include data on natural and laboratory-associated infections of man and virologic observations of humans.

**Table 1.** Major viral hemorrhagic fevers (VHF)

| Family/Genus | Disease | Geography | Laboratory biosafety level | Inter-human transmissibility |
|---|---|---|---|---|
| **Arenaviridae** | | | | |
| *Lassa virus* | Lassa fever | West Africa | BSL-4 | Person-to-person Rare nosocomial |
| Junin, Machupo, Guanarito, and Sabia viruses | Various South American HFs | Argentina, Bolivia, Venezuela, Brazil | BSL-4 | Rare nosocomial Rare inter-human |
| **Bunyaviridae** | | | | |
| *Phlebovirus* | Rift Valley fever | Sub-Saharan Africa | BSL-3[a]/BSL-4 | None |
| *Nairovirus* | Crimean-Congo HF | Africa, Asia, Balkans | BSL-4 | Occasional nosocomial |
| *Hantavirus* | HF with renal syndrome | Europe, Asia, perhaps elsewhere | BSL-3[b] | None |
| | Hantavirus pulmonary syndrome | Americas | BSL-3[b] | None |
| **Filoviridae** | Marburg and Ebola HF | Sub-Saharan Africa | BSL-4 | 5–25% in unprotected patient care and household setting |
| **Flaviviridae** | Yellow fever | Tropical Africa and South America | BSL-3[a] | None |
| | Dengue HF/dengue shock syndrome | Tropics | BSL-2 | None |

[a] Laboratory workers must be protected by vaccine. In the case of Rift Valley fever virus, vaccinated workers use BSL-3 precautions within the laboratory, but additional precautions are required to prevent escape of the virus into the environment
[b] Work with the virus can be carried out at BSL-3, but special precautions are required for inoculation of natural reservoir rodents

*Field observations*

Field data are necessarily limited in their reliability. They often are collected in areas of the world with an underdeveloped public health and economic infrastructure. The virus concerned may be endemic, thus masking person-to-person transmission which may be attributed to other sources of infection; alternately, previous exposure to the virus may have provided immunologic protection to medical staff. Occult use of unsterilized needles and other instruments can lead to iatrogenic transmission that can be mistaken for other mechanisms of spread. The worst outbreaks are more likely to be reported, but small episodes can easily be overlooked or mis-attributed. In spite of these caveats, several observations have been made that are worth consideration.

Lassa fever can be spread person-to-person among medical staff in the absence of parenteral exposure [58], but hospitals in Sierra Leone have reported a very low incidence of disease among health care providers in spite of caring for large numbers of Lassa fever patients [32]. However, there has been at least one large, well-documented nosocomial Lassa fever outbreak in Nigeria in which 24 patients were infected, with a 54% mortality rate [10]. Careful investigation failed to identify the mode of transmission. The index patient was severely ill with prominent cough and her bed was in front of an open window that ventilated the general ward, suggesting the possibility of aerosol spread. Investigation of a Nigerian family outbreak in 1993 (P. E. Rollin, unpubl. data) also suggested person-to-person spread without any parenteral component. Seven Lassa fever patients have been cared for in Western countries without incident [24].

Argentine HF has been suspected of nosocomial spread in a single patient who disseminated the disease to family and medical staff (J. I. Maiztegui, unpubl. data). Bolivian HF was generally regarded as a disease with little potential for person-to-person spread, but two spouses of infected investigators became ill outside the disease-endemic area [20]. A serious nosocomial Bolivian HF outbreak occurred in 1971 outside the disease-endemic area; family members were in close contact with the index patient case including demonstrative embraces, but one student nurse was infected after only watching a demonstration of changing bed linen by a nursing instructor [68]. In 1994 a Bolivian HF case in a private home was associated with subsequent fatal infection among 4 of 6 exposed household members and 2 additional non-household family members under circumstances suggesting that at least some were likely small particle aerosol mediated (T. G. Ksiazek, unpubl. data).

Several nosocomial Crimean-Congo HF outbreaks have been noted. These are usually associated with an index patient who presents with severe illness and hemorrhage and considerable exposure to blood occurs during patient care procedures. The majority of cases are not associated with inter-human transmission [81–83].

Among filovirus outbreaks, inter-human transmission is the rule and medical staff are particularly at risk. Most, but not all, infections involve close contact with sick patients. In modern health care settings, Marburg virus has resulted in a low attack rate among hospital staff. In less well-appointed hospitals, Ebola virus has devastated the medical staff and led to the abandonment of hospitals [14, 33, 62, 85]. About 5% of family members were infected during the 1976 Ebola epidemic in Zaire [33]. During the 1995 Zaire Ebola epidemic a 16% secondary attack rate among family members was found and two patients were noted to have particularly large numbers of infected contacts [15]. In all cases, epidemics have been controlled by instituting proper medical care precautions in which mask, gown, and gloves are worn and by educating local residents about ways to reduce their risks, including changes in burial practices and minimizing contact with sick persons.

These patterns strongly suggest that the major modes of inter-human transmission of HF viruses are by direct contact, larger droplets, or fomites and that virus spread requires close proximity. However, it is difficult to exclude a component of small particle aerosol transmissibility, particularly in the hospital-centered outbreaks. The reasons for these occasional common-source epidemics are unknown. One possibility is that a uniquely pathogenic or transmissible virus strain is circulating. A more likely explanation is that virus replication occurs at an unusually high level in a particular patient and this is responsible both for the severity of the patient's illness as typically observed and for an increased shedding of virus into the environment, resulting in a cluster of secondary cases. It is instructive to compare the situation now prevailing in VHF epidemiology with Langmuir's reprise of concepts of measles, smallpox, and rubella epidemiology [46]. The similarities among these three diseases include the incomplete understanding of transmission patterns if a component of aerosol transmission is not considered, the reluctance of many to accept invisible aerosols as being an important vehicle of infection, and the presence of occasional "super-spreaders." It is not known what the eventual explanation will be in the VHF.

### Natural and laboratory infections

In the case of arenaviruses, rodent-to-human transmission frequently results from human exposure to aerosolized excreta or secreta of infected rodents. Arenaviruses, Rift Valley fever virus, and to a lesser extent Crimean-Congo HF viruses are well known for their small particle aerosol transmissibility in the laboratory [2, 72]. There is much less experience with filoviruses because they have been BSL-4 agents virtually since their initial isolation. In the original Marburg virus outbreak, numerous laboratory staff working with monkey blood and tissues as well as cell cultures were infected, but none of these persons were using appropriate precautions and they were exposed to large amounts of virus-containing fluids.

### Virus in blood and body fluids

Another relevant variable for the transmissibility of the VHF agents, by whatever means, is the amount of virus present in blood and other patient materials (Table 2). Data are incomplete, but suggest considerable variation among

**Table 2.** Viremia and excretion of virus in VHF

| Disease | Site | Infectivity | Last positive result from date of onset | Reference |
|---|---|---|---|---|
| Lassa fever | Blood | $2.0–8.1 \log_{10} TCID_{50}/ml$ | >16 days | 9, 40, 59 |
| | Throat swab | $1–2.3 \log_{10} TCID_{50}/ml$ | 13–24 days | 9, 40, 59 |
| | Urine | $1–2 \log_{10} TCID_{50}/ml$ | 32 days | 9, 40, 59 |
| | Cerebrospinal fluid | $1.6–3.1 \log_{10} TCID_{50}/ml$ | 5–14 days | 18, 40 |
| | Pleural fluid | $2.25–>4.5 \log_{10} TCID_{50}/ml$ | 13–14 days | 9, 59 |
| Argentine and Bolivian HF | Blood | $2.3–5.2 \log_{10} PFU/ml$ | Day 13 | 68 |
| | Throat swab | $2.8 \log_{10} PFU/ml$ | Day 10 | 31, 68 |
| | Urine | Positive, rare | Day 8 | 31 |
| | Breast milk | Positive | | 49 |
| Rift Valley fever | Blood | $4.2–8.1 \log_{10} MIC LD_{50}/ml$ | Acute blood | 55 |
| Crimean-Congo HF | Blood | upto $6.2 \log_{10} TCID_{50}/ml$ | Day 9 | 73, 82 |
| | Throat swab | Negative | | 73 |
| Marburg HF | Blood | Positive | Day 7 | 28, 80 |
| | Throat swab | Positive | Unknown | 28 |
| | Urine | Positive | Unknown | 28 |
| | Feces | Positive | Unknown | 28 |
| | Semen | Positive | 83 days | 50 |
| | Ant. Ocular chamber | Positive | 80 days; neg. 10 weeks later | 28 |
| Ebola HF | Blood | $4.5–6.5 \log_{10} PFU/ml$ | 8 days | 33 |
| | Semen | Positive | 61 but not 74 days | 21 |

$TCID_{50}$ = 50% tissue culture infectious dose
PFU = Plaque forming units

viruses. The lack of data on viral titers and duration of viral secretion in many bodily secretions such as faeces, urine, saliva, semen, cervical fluids, sweat, and tears underscores the difficulties in formulating guidelines for managing convalescent patients.

## Experimental infections

Because of the lack of detailed information about human infections, it is worthwhile to examine the lessons that animal models may contribute. Models that mimic the pathogenesis of human diseases in most respects are selected for consideration, although there may be quantitative differences compared with diseases in humans. These models have been invaluable in developing diagnostic and therapeutic approaches to the various forms of VHF. In this discussion, these models may also provide the only data on the early stages of infection, which have never been studied in humans.

### Pathogenesis of virus infections (Fenner model)

Our understanding of virus pathogenesis is often measured against the classic model developed by Frank Fenner for mousepox or ectromelia infection of the laboratory mouse (Fig. 1) [23]. While the basic scheme has many variations, the conceptual pattern holds for most acute viral infections. Mice are inoculated with ectromelia virus subcutaneously in the hind limb and virus is carried in the lymph to the draining lymph node. It multiplies locally and is discharged via efferent lymphatics into the subclavian vein, resulting in what is referred to as the "primary viremia." The primary viremia infects certain target organs that are important sites of multiplication for virus and seed the bloodstream with the "secondary viremia." This viremia then results in infection of the skin and other organs that serve as portals to the environment.

### Pathogenesis of hemorrhagic fever models

Although the detailed pathogenesis of most of the VHFs has not been described, excellent animal models exist for several of these diseases and patterns resembling Fenner's model are commonly found. Infection with arenaviruses, Rift Valley fever virus, or filoviruses by the parenteral route results in viremia and subsequent infection of target organs. The timing of infection of organs such as lung, salivary gland, kidney, bladder, and others that can lead to excretion of virus to the outside is not known in detail, but all of these are targets for one or more of these viruses. In several models, the actual excretion of virus has been measured (Table 3). As expected from the Fennerian paradigm, the excretion of virus follows the onset of viremia and overt disease.

Lassa viruses infect macaques by most routes of inoculation; in the few instances in which infectivity has been measured, the severity of disease reflects virulence for humans. Highly virulent strains are almost uniformly fatal for cynomolgus macaques [37]. The disease is characterized by a viremia that is first

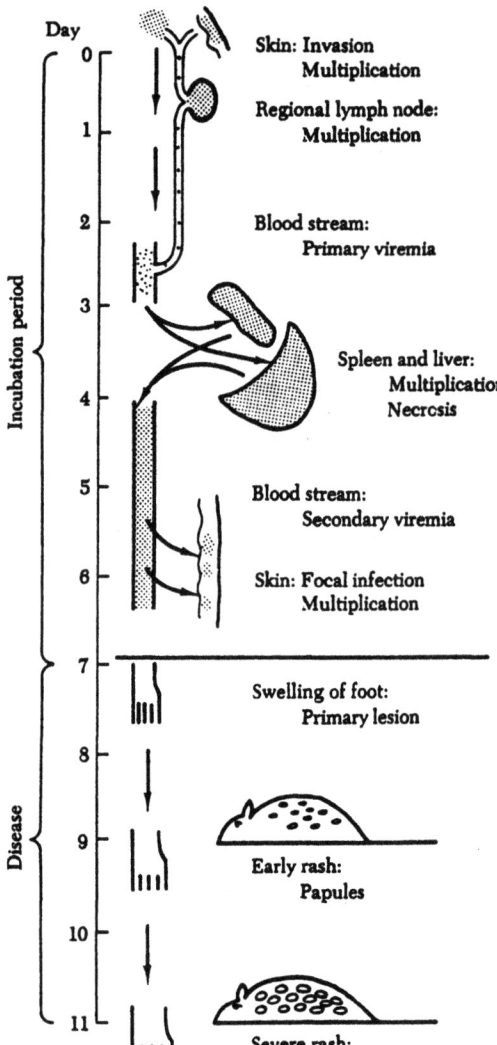

**Day**

Incubation period

0 — Skin: Invasion
        Multiplication

1 — Regional lymph node:
        Multiplication

2 — Blood stream:
        Primary viremia

3 —

4 — Spleen and liver:
        Multiplication
        Necrosis

5 — Blood stream:
        Secondary viremia

6 — Skin: Focal infection
        Multiplication

Disease

7 — Swelling of foot:
        Primary lesion

8 —

9 — Early rash:
        Papules

10 —

11 — Severe rash:
         Ulceration

**Fig. 1.** Diagram illustrating the pathogenesis of mousepox. Ectromelia virus gains entry through minute abrasions of the skin in which it multiplies to produce a primary lesion. While this lesion is developing, a series of invasive steps produce a secondary viremia that seeds the skin and other organs with virus. The rash appears about 3 days after the primary lesion (reproduced from [23])

detectable 3 days after intramuscular inoculation, increases to a peak concentration $> 6 \log_{10} \mathrm{PFU/ml}$ around day 10, and continues until the animals die around day 14–18. By the time of death, the virus is actively replicating in most organs, including the urinary tract epithelium and salivary gland. Thus, throat swabs contain virus after day 10 until death. Significantly, virus occasionally has been isolated from throat swabs of animals whose viremia was suppressed by treatment with ribavirin (P. B. Jahrling, unpubl. data). Lassa virus concentrations in urine are quite variable, but when efforts are made to stabilize viral infectivity by addition of protein, viruria titers usually range from 5.5 to $6.5 \log_{10} \mathrm{PFU/ml}$ beginning on day 10 and persisting until death. Monkeys inoculated with Bolivian and Argentine HF viruses (Machupo and Junin viruses, respectively) live somewhat longer but exhibit similar patterns. For the hemorrhagic Espindola strain of Junin, viremias range from 6   7 $\log_{10} \mathrm{PFU/ml}$, peaking

**Table 3.** Viremia and excretion of virus in primate models of VHF

| Disease | Site | Infectivity (log₁₀ PFU/ml) | Peak day | Duration (range in days) | Comments | Reference |
|---|---|---|---|---|---|---|
| Lassa | Blood | 5–6 | 10 | 3–18 | Strain-dependent Positive while on ribavirin and viremia undetectable. | |
| | Throat swab | 2–3 | 12 | 10–18 | | |
| | Urine | 5.5–6.5 | | 10–18 | Highly variable (need protein to stabilize infectivity?) | |
| | Feces | NT | | | | |
| | Semen | NT | | | | |
| Argentine HF | Blood | 6–7 Espindola strain | 17 | 10–25+ | Declining at death. | 52, 53 |
| | | 2–3.1 Ledesma strain | 10 | 10–18 | | |
| | Throat swab | 4–6 (Espindola) | 21 | 10–25+ | Throat swab titer often > serum viremia. | 52, 53 |
| | | 2–4 (Ledesma) | 14 | 10–25 | | |
| | Urine | NT | | | | |
| | Feces | NT | | | | |
| | Semen | NT | | | | |
| Rift Valley fever | Blood | 5.5–6.5 | 2 | 1–8 | 3–4 day duration, longer in fatal infection. Viremia peak does not correlate with disease severity. | 65 |
| | Throat swab | Occasional positive | | | Epistaxis? | 65 |
| | Urine | NT | | | | |
| | Feces | NT | | | | |
| | Semen | NT | | | | |
| Crimean-Congo HF | Blood | No model | | | | |
| Marburg | Blood | 6–7 | 7 | 3–21 | | 75 |
| | Throat swab | 0.7; "positive" | 7 | 3–11 | | 3, 71; 7 |
| | Urine | 1.5; 6 | 9 | 7–11 | | 5, 76 |
| | Feces | NT | | | | 3, 71; 7, 4 |
| | Semen | NT | | | | |
| Ebola | Blood | 5–7 | 7 | 3–11 | | 26 |
| | Throat swab | 2–4 | | 8–14 | | 26 |
| | Urine | "trace" | terminal | | | 26 |
| | Feces | 2–3 (anal mucosa) | terminal | | Haemorrhage? | 26 |
| | Semen | NT | | | | |

*NT* Not tested
Espindola is a haemorrhagic strain of AHF
Ledesma is a neurolic strain of AHF

around day 17 and declining until death around day 25. The neurologic Ledesma strain is associated with lower viremias. Both viral strains are shed in the saliva to a greater extent than thought to occur in humans; peak titers occur after peak viremias, but persist until death, often at concentrations higher than the corresponding viremias. No attempts have been reported to isolate virus from faeces or semen.

In contrast with the VHFs associated with the arenaviruses, Rift Valley fever (RVF) virus is not routinely fatal for macaques. However, RVF did cause a significant viremia ($5.5$–$6.5 \log_{10}$ PFU/ml) which peaked early (days 2–3) before subsiding. Severity of disease did not correlate closely with magnitude of viremia, but viremia that lasted longer than 3–4 days did. Although attempts to isolate virus from throat swabs have not been reported, the occurrence of epistaxis in acutely ill monkeys probably means that oropharyngeal secretions are infectious. Like RVF, yellow fever infection of macaques is characterized by high viremias which rapidly resolve. Yellow fever infection of rhesus macaques is associated with 50% mortality; lethally infected monkeys develop viremias in excess of $7 \log_{10}$ PFU/ml in contrast with survivors with peak viremias of 4.5 to $5.2 \log_{10}$ PFU/ml. Despite systematic testing, yellow fever virus has not been isolated from the throat swabs. This correlates with the lack of ready yellow fever transmission in humans.

Despite the fulminant nature of human Crimean-Congo HF infection, no adequate animal model is available for pathogenesis studies.

Marburg virus infections are usually fatal for macaques exposed by any route. Time to death depends on viral strain and dose; generally, monkeys die 7–11 days after exposure with viremias developing early and reaching peak titers around $7 \log_{10}$ PFU/ml by day 7. Throat swabs become positive for virus about the same time, but quantitative data are lacking.

Likewise, urine frequently contains Marburg virus soon after onset of viremia, but titers are highly variable [3, 75]. Experimental Ebola virus infections are similar, with infectious viral burdens dependent on viral strain and route of exposure [25, 26, 36]. Most macaques inoculated with Ebola virus develop lethal infections and high viremias before death. Virus titers in throat swab range from 2 to $4 \log_{10}$ PFU/ml, and urine specimens contain trace amounts of virus before death. Likewise, swabs of anal mucosa of terminally ill monkeys frequently contain infectious virus, which may reflect upper or lower gastrointestinal bleeding [36]. Recent examination of skin biopsy specimens from Ebola virus-infected monkeys by using immunohistochemical methods have demonstrated viral antigens in skin and appendages (N. K. Jaax, unpubl. data), analogous to the observations in human patients (S. Zaki, unpubl. data).

### Experimental transmission

Transmission between infected laboratory animals may occur in the course of other experimental studies or exposure of naive animals may be designed as a controlled experiment. Inadvertent transmission of Lassa virus to monkeys was reported to occur when 12 cynomolgus macaques were introduced into a room occupied by guinea pigs and mice inoculated 12 days previously with a virulent strain of Lassa virus. Before the aerosol transmissibility of Lassa virus was fully appreciated, monkeys were conditioned to the BL-4 laboratory shared with other animals. Although the monkeys were not inoculated and had not been handled in any way, within 3 weeks, two of them died and Lassa virus was isolated from all organs tested. At other times, uninoculated monkeys were observed to seroconvert to Lassa virus without developing overt disease. While transmission between monkeys is possible, documented occurrences of inadvertent Lassa infection in monkeys have always involved exposure in rooms containing infected guinea pigs or mice and their bedding (P. B. Jahrling, unpubl. obs.). However, infectious Lassa virus titers in monkey urine frequently exceed $6 \log_{10}$ PFU/ml, sufficient to transmit infection either by droplets or by true aerosol.

Transmission of Junin virus was also documented, on two occasions, to occur between inoculated monkeys and controls in adjacent cages. In both instances, this involved the Espindola strain, which is associated with the hemorrhagic form of the disease, characterized by mucous membrane hemorrhage coupled with high viremia and high titers of virus in throat swab, salivary gland, and urinary tract tissues (McKee, Peters, and Mahlandt, unpubl. obs.).

Experimental transmission of Marburg virus was reported [71, 75] to occur most efficiently when monkeys were co-housed. Monkeys caged separately but downwind of infected monkeys also became infected. In addition, transmission of Marburg virus occurred between monkeys housed in special cages where monkeys faced each other nose-to-nose and contact with feces and urine was minimized by physical barriers. The latter experiment does not exclude the possibility of transmission through larger droplets on the conjunctiva or through ingestion.

Both guinea pigs and monkeys have been experimentally infected with Ebola virus by direct instillation of drops onto the eye or into the throat ([34], P. B. Jahrling, unpubl. data). This observation raises the importance of full face protection for health care workers to avoid infection via the conjunctival or alimentary routes. As with many of the other VHF agents, Ebola virus has been documented to spread between experimentaly infected and control monkeys in separate cages [35]. In addition, during the Ebola epizootic at Reston, there was clear evidence of transmission within and between rooms of monkeys under conditions in which aerogenic spread was considered possible [19]. This finding is compatible with the observation that Ebola viral antigen in the lungs of the infected monkeys had a bronchocentric distribution described previously for monkeys experimentally exposed to Ebola virus aerosols in a controlled experiment [39]. This mode of transmission also fits with the demonstration of Ebola viral antigens in secretions on mucosal surfaces of the nose and oropharynx [29, 36] and respiratory epithelium of experimentally infected monkeys and guinea pigs (Fig. 2) (P. B. Jahrling, unpubl. data). It may be that infection of respiratory epithelium leads to a vicious cycle in which increased viral shedding

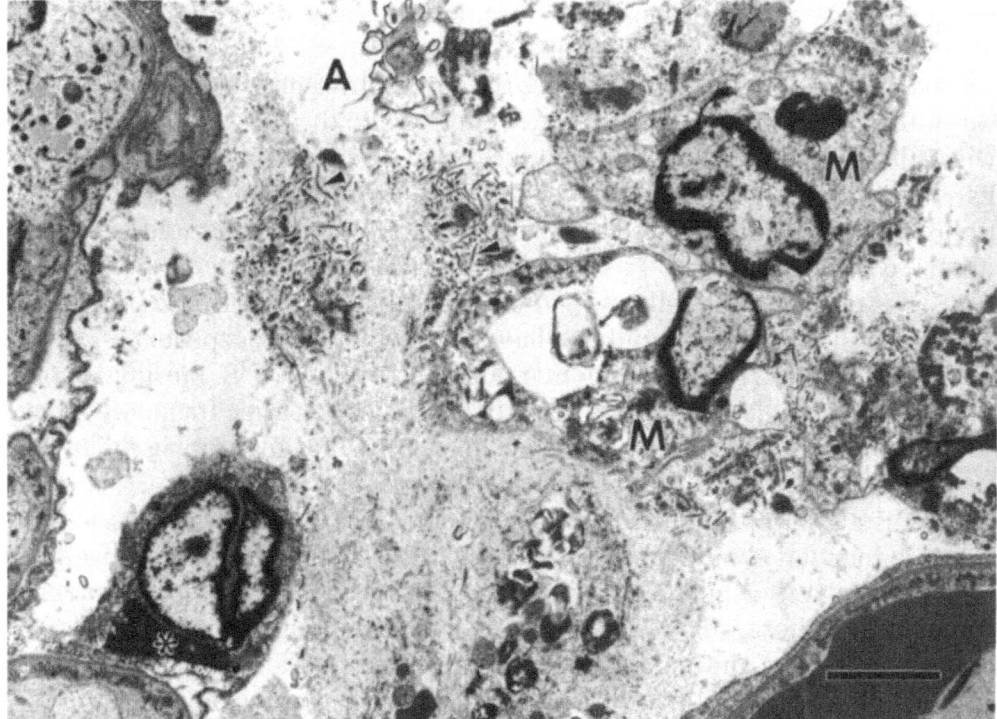

**Fig. 2.** Photomicrograph of lung tissue from an orally infected guinea pig showing cluster of Ebola virus (arrowheads) particles free in alveolar space (*A*) associated with degenerate alveolar macrophages (*M*). Note the characteristic filoviral inclusion (*) in the cytoplasm of a type I alveolar epithelial cell. Bar: 3.0 μm. Kindly provided by Thomas W. Geisbert

via the respiratory route results in increased secondary infection via the respiratory route [12, 19].

## Small-particle aerosols

### *Definition*

When liquid suspensions containing viruses are dispersed in the air, the sizes of the particles formed are extremely important in their subsequent behaviour (Fig. 3). Larger droplets (> 5–10 μm) settle out rapidly and under ordinary circumstances travel no further than one or perhaps 3 meters, during which they may contact skin or mucous membranes and their presence may actually be felt as moisture or another sensation. If air containing such droplets is inspired, they are largely trapped in the turbinates or impinge on the posterior pharynx.

Smaller particles (1–5 μm) have markedly different properties and will be referred to here as small-particle aerosols or simply aerosols. They will remain dispersed in the air and circulate as airborne particles for long distances without settling. If inspired, they will penetrate deep into the lung, and Brownian motion, sedimentation, and turbulence will result in retention of about half of them in the

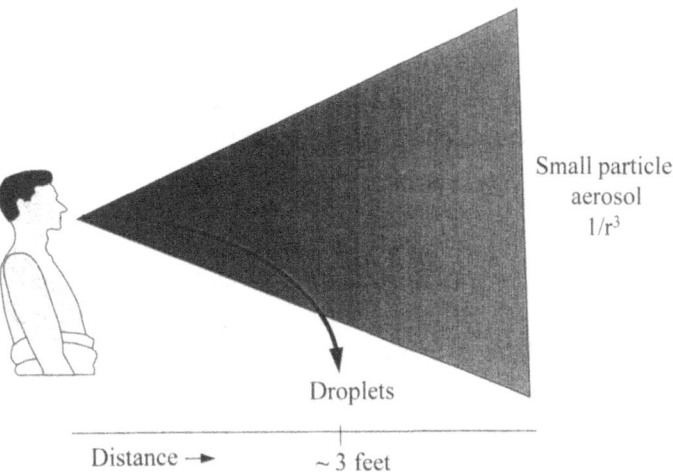

**Fig. 3.** Hypothetical fate in the environment of droplets and 1–5 μm small-particle aerosols generated from an infected patient. Small-particle aerosols do not settle out at an appreciable rate but spread, so that as the distance from the source increases, their concentration in air decreases in proportion to $1/r^3$. This calculation ignores decay of the aerosol from physical factors, such as drying and ultraviolet radiation or convective disturbances by air currents (biological $t_{1/2}$ for many VHF agents is approximately 15–30 mins)

lower respiratory passages or alveolae [61]. These particles are not efficiently filtered from air by ordinary surgical masks and therefore enhanced respiratory protection must be used. Smaller particles ($< 1$ μm) will remain in suspension and be carried on air currents, but their pulmonary retention is much less until they reach very small sizes and retention increases again.

Regardless of particle size and the major anatomic area of deposition, some particles will be retained at all sites. Particle handling varies with respiratory rate, mouth-breathing, and other factors to make these generalizations very approximate. Furthermore, ciliary action will move materials through the respiratory tract, including the oropharynx where they may be swallowed.

Thus, the generation of infectious small particle aerosols raises several concerns: 1) they may be carried on air currents and may pose a hazard if inspired, 2) special filters must be used for respiratory protection against them, and 3) they may contain agents that have a different pathogenesis because of the site of infection, as seen for the non-viral pathogens *Bacillus anthracis* (anthrax) and *Yersinia pestis* (plague).

### Aerosol infectivity of viruses

Some microorganisms are infectious when aerosols containing them are generated, whereas others pose little infectious threat in this regard. Several well-known viral diseases are transmitted efficiently by small-particle aerosols, including influenza, measles, and rubella. The continued maintenance of such viruses in nature implies that they are present in external secretions, that energy-requiring processes such as coughing generate infective small-particle aerosols, and that the viruses are sufficiently stable in aerosol to persist until inspired. These properties and their aerosol infectivity result in continuous transmission of the viruses between suspectable hosts. The VHF viruses are zoonotic, however, and have other mechanisms for their persistence in nature [69]. Thus, in spite of the aerosol infectivity of these viruses, they are not continuously transmitted between humans by the aerosol route.

Aerosol infectivity of viruses causing HF can be suspected in some cases, arenaviruses and hantaviruses in particular, because the natural spread to man is by the aerosol route. Other viruses, such as Rift Valley fever and yellow fever viruses, have substantial aerosol infectivity but their natural mode of transmission is by arthropod vectors. In some cases, the infectivity of these viruses has been formally measured (Table 4). As can be seen, the infectivity of viruses carried by rodent hosts ranges from 1 to 2 000 PFU inhaled; in some cases, the outcome in some is measured by infection only, but in others lethality is the end-point. Data for non-human primates are less complete; because of the hazards and expense in conducting such studies, it was often only confirmed that aerosols were infectious without seeking an end-point. When parenteral infectivity is compared with the effective aerosol dose, the former is usually more efficient.

The pathogenesis of aerosol infections resembles that of parenteral infection, with perhaps somewhat more infectivity measured in lung homogenates after aerosol infection (Fig. 4) [7, 42, 78, 79]. The major exception to this general finding has been described in reports of Ebola virus experiments. Aerosol infections with the Ebola Zaire subtype have led to more pulmonary involvement than that which occurs with parenteral infection [34]. This pattern parallels the clinical observations on monkeys undergoing transmission of the Reston subtype in which successive generations of animals clinically evidenced more pulmonary secretions and upper respiratory involvement; these findings have not been studied systematically to confirm them [66].

### Generation and stability of aerosols

Particles in the critical 1–5 µm range require energy to be produced from bulk fluid. This energy can vary in efficiency and source from spray nozzles, rapid turbulent air flow of a cough, or conversion of kinetic energy by fluid falling on a hard surface. Once formed, an aerosol can decay physically by the actual loss of particles or by the loss of infectivity independent of the physical loss of particles. The decay will depend on a number of factors, including relative humidity and the composition of the fluid in which the virus was suspended. In addition, two major factors will decrease the concentration of aerosols in natural situations: ultraviolet light from the sun and dilution by diffusion and wind currents.

Measurements of the aerosol stability of VHF agents made under controlled conditions in the laboratory show a significant stability, with the half-time for loss of infectivity usually ranging from 10 min to 1 h (Table 4). As with most lipid-enveloped viruses, the greatest stability is at the lowest values of relative humidity.

**Table 4.** Aerosol infectivity of some haemorrhagic fever viruses

| Virus | Non-Human Primate Dose/$LD_{50}$ | Other species dose | Decay ($T_{1/2}$ in min) | Reference |
|---|---|---|---|---|
| Lassa | <465 PFU[a] | $ID_{50}$ = 15 PFU guinea pig | 10–55 | 79 |
| | | $LD_{50}$ = 1995 PFU guinea pigs (Hartley) | 13–35 | 4 |
| | | $LD_{50}$ = 10 PFU guinea pigs (strain 13) | | 4 |
| Junin | <50 PFU | $LD_{50}$ = <350 PFU guinea pig (strain 13) | 16–28 | 42, 78 |
| | | $LD_{50}$ = 3 PFU guinea pig | 12–17 | 4 |
| Rift Valley fever | <76 $MIPLD_{50}$[b] | $LD_{50}$ = 0.525 $MIPLD_{50}$ hamster | 19–61 | 56 |
| | | $LD_{50}$ = 79–398 PFU mouse | | 7 |
| | | $LD_{50}$ = 10 PFU rat (Wistar-Furth) | | 6 |
| | | $ID_{50}$ = 5–7 $MICLD_{50}$ kitten | | 41 |
| | | $ID_{50}$ = ~ 25 $MICLD_{50}$ puppy | | 41 |
| Hantaan | | $ID_{50}$ = 0.5 PFU rat | | 63 |
| Marburg | 1.3 PFU | $LD_{50}$ = 0.9 PFU guinea pig | | 3 |
| Ebola | <400 PFU | | | 39 |
| Yellow fever | <6 $MICLD_{50}$[c] | | 15 | 56 |

$LD_{50}$ (50% lethal dose) & $ID_{50}$ (50% infectious dose); dose required to detect 50% of specified outcome
[a]Infectivity measured by plaque forming units (PFU) in cell culture monolayer
[b]Infectivity measured by 50% mouse intraperitoneal lethal dose ($MIPLD_{50}$)
[c]Infectivity measured by 50% suckling mouse intracranial lethal dose ($MICLD_{50}$)

It would be desirable but obviously difficult to measure the quantity of aerosolized particles emanating from infected animals or patients. In addition to methodologic issues, particles may decrease in size due to drying (so called "droplet nuclei") or dynamically hydrate in the respiratory tract [43, 61]. Attempts to quantify aerosols from coughs by balloon collections have suggested that a single representative cough by a Coxsackie virus A-21-infected person might generate $2.5 \times 10^4$ particles between 1–8 μm in diameter with a total volume of $1.5 \times 10^{-7}$ ml [30]. A "typical" sneeze resulted in about 30 times the volume of particles as a cough, but when virus was assayed, a cough was as likely to result in recovery of airborne virus. Nevertheless, small quantities of virus ($50–500$ $TCID_{50}$) were recovered from room air and this was sufficient to infect volunteers exposed across a double wire screen 4 feet wide [17, 43]. The aerosol infectious dose of Coxsackie A-21 virus is 28 $TCID_{50}$ for humans, and the concentrations in respiratory secretions have been reported as 30 to 1 000 $TCID_{50}$ [17] and occasional titers as high as $10^6$ (Couch, pers. comm.).

In the laboratory setting, common mistakes in technique can result in aerosolization of $10^{-4}$ to $10^{-7}$ ml of sample in a respirable particle size range. Centrifuges can impart the energy to produce much larger aerosols [54], and these aerosols have been responsible for numerous laboratory infections with some of the viruses under discussion here. Only a few of the manipulations that commonly occur in the patient care setting have been evaluated for their potential to generate aerosols. Studies during hemodialysis or dental procedures failed to detect any significant aerosolization of hepatitis B antigen or hemoglobin markers [70], although some surgical power tools may produce airborne particles of hemoglobin-containing fluid [38]. There is no significant indication of aerosol infectivity of the traditional blood-borne pathogens, such as hepatitis B virus or human immunodeficiency virus (HIV), either in the patient care or laboratory setting.

### Theoretical assessment of aerosols in nosocomial VHF transmission

The possibility of aerosol transmission can be approached by considering the quantitative aspects of the generation of aerosols and the infectious process. As discussed above, the concentration of virus in body fluids will virtually always be $< 10^7$ infectious units/ml and may be several orders of magnitude less. Thus, a patient with a very high virus titer in pulmonary secretions may aerosolize only a single infectious unit or even less with a cough. More vigorous situations, such as arterial bleeding or suctioning of pulmonary secretions, would still be expected to result in low concentrations of true small-particle aerosols.

Of course, these theoretical considerations can only be used to give a crude indication of the possibility of transmission. If the data for experimental animals shown in Table 4 give some idea of the infectivity for man, then the low-intensity aerosols expected from information on virus titers and aerosol generation would suggest that there is a definite but low risk of aerosol transmission. Another consequence of the low concentration of 1–5 µm particles is that the risk of transmission at a distance is greatly lessened. In addition to decay and wind currents, the basic physical nature of an aerosol dictates that the concentration will decrease as the third power of the distance diffused (Fig. 3). Thus, as the distance from a source doubles, the concentration of the aerosol declines 8-fold. This physical fact also makes it difficult to distinguish between epidemiologic evidence of droplets and aerosol transmission since the aerosol is much stronger closer to the patient, where droplets and fomites are also most important. The distinction may nevertheless be important because the smaller particles readily bypass and pass through a surgical mask.

## Synthesis for individual viruses

Arenavirus-infected patients usually pose only a small risk for family members or health-care providers, although Lassa fever may often result in secondary cases among close contacts. Rarely there have been instances of dissemination by individual patients under circumstances that suggest an element of small particle aerosol transmission, but even these episodes have resulted in low attack rates among those exposed. The ability of arenaviruses to infect a variety of rodents other than their reservoir species with transient shedding makes it prudent to exclude the possibility of infection of local rodents. It is unlikely that arenaviruses could become established in local species, but there are troubling examples to the contrary. For instance, lymphocytic choriomeningitis virus can be maintained in hamster colonies and Lassa virus can cause chronic viremia in genetically susceptible mice [64].

Rift Valley fever has never resulted in well-documented secondary human cases in spite of its amply-proven hazard in the virology laboratory. Care should be taken to exclude contact with mosquitoes because of the wide variety of arthropods in North America and elsewhere that may become orally infected and transmit the virus. RVF virus also has the widest mammalian host range for acute infections and the greatest danger of establishing itself as an imported veterinary pathogen.

Crimean-Congo HF has infrequently but regularly caused nosocomial epidemics but the modes of transmission have never been clearly defined. The propensity for hemorrhage among many of these patients enhances the potential for exposure to blood, and it seems likely that the majority of the subjects were infected by direct contact, larger droplets, or indirect contact via contaminated fomites and biological fluids.

Filoviruses are notorious for person-to-person transmission among family and medical staff, but generally in circumstances involving close exposure and often contact with blood itself. No epidemiological evidence exists for small particle aerosol transmission from person to person. Because of the large knowledge gaps concerning the pathogenesis, genetic variation, and natural reservoirs of this virus family, it continues to be the most troublesome of all the viruses causing HF.

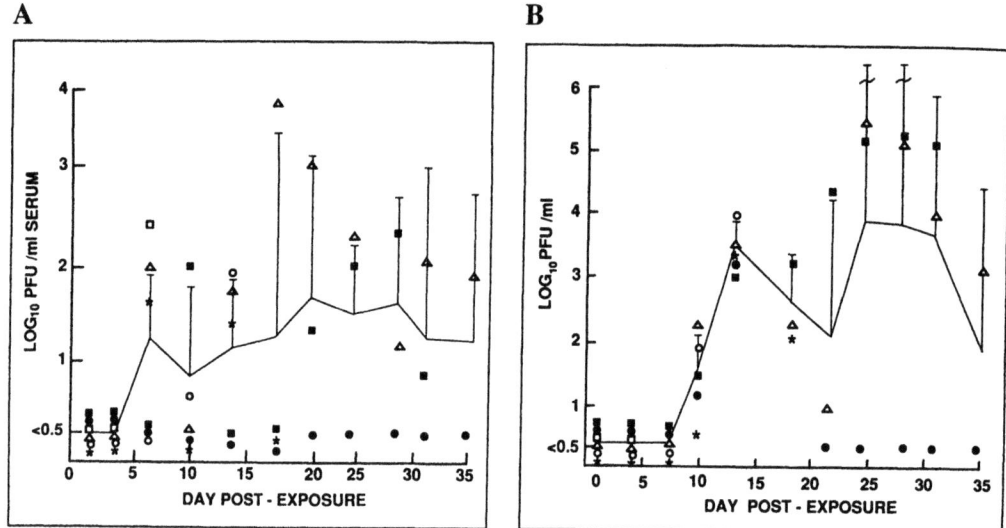

**Fig. 4.** Viremia and virus in throat swabs from monkeys experimentally infected on day 0 with Junin virus by aerosol exposure. Rhesus monkeys in the high-dose group were presented with $10^{4.3}$, $10^{3.9}$, and $10^{4.3}$ PFU of virus. Rhesus monkeys in the low-dose group were presented with $10^{1.9}$, $10^{1.6}$, and $10^{1.7}$ PFU of virus (reproduced from [42]). **A** Serum viremia in rhesus monkeys infected by an aerosol. Low dose: ● = 121N [animal number], ★ = B52, ■ = T280; high dose: △ = P829, ○ = 556, □ = B4. Geometric mean values are connected by a straight line, and the upper standard errors are shown. **B** Titers of Junin virus in oropharyngeal swabs from monkeys infected by aerosol. Low dose: ● = 121N, ★ = B52, ■ = T280; high dose △ = P829, ○ = 556, □ = B4. Geometric mean values are connected by a straight line, and the upper standard errors are shown

Yellow fever virus is a highly infectious pathogen that caused numerous laboratory infections before effective vaccination became available. Patients are usually only viremic for the first 4 days of illness so that by the time substantial virus-induced target organ damage has occurred, they provide no appreciable risk except for residual virus in viscera. Early in disease, the risk for transmission to the ubiquitous urban mosquito vector *Aedes aegypti* is high.

## An approach to patient care

### Spectrum of patient containment

In disease-endemic areas, most patients with VHF are cared for without special precautions, and mask, gown, and even gloves are usually not available or not used. In this setting, the risk of transmission to care-givers is usually low; however, occasional secondary cases of Lassa fever and other diseases occur among care-givers just as they do among family members. In the case of filoviruses, devastating, progressive, hospital-based epidemics can occur. In the most recent outbreak in Zaire in 1995, most cases were community acquired but 24% were in medical personnel, and hospitals played a key role in amplification as well as control [14, 15, 62]. Occasionally, exceptional but severe, acute clusters of VHF have been observed that seem to be related to a single case with unusual characteristics such as reported with Crimean-Congo HF, Ebola, Lassa, and Machupo virus infections for which the mode(s) of transmission is not well-defined ([10, 68], A. S. Khan, unpubl. data). None of these outcomes would be regarded as acceptable if VHF were exported to non-endemic areas of the world.

At the other end of the spectrum is virtually complete containment of the patient with BSL-4 level precautions [45]. The patient can be isolated in a room with redundant systems to provide assurance of negative air pressure, filtered air entering and leaving, pass-through autoclaves for exit of used materials, and sterilization of all effluent sewage. Care would be provided by specially-trained staff garbed in impermeable suits under positive air pressure and who exit through a disinfectant shower. These systems were reasonable 25 years ago when they were put in place and still provide a useful backup for future eventualities. Current experience shows that these precautions may be used but suggests they are not needed for care of patients infected with any known pathogen.

Several intermediate levels of precaution can reasonably be applied. Clearly, a surgical mask with a visor or separate goggles, gown, and gloves should be used routinely. The degree of protection for the individual and the environment can be increased by providing impermeable gowns or garments which are removed prior to leaving the contaminated area. These can be worn over scrub suits, which would remain in the anteroom to the contaminated area.

There is also an opportunity to provide more efficient protection against inspired particles for patients who have a prominent cough, vomiting, diarrhea, or hemorrhage that range from disposable dust-mist or N95 respirators, through fitted high efficiency particulate (HEPA or N100) respirators, and finally powered air purifying respirators (PAPR). The latter is commercially available as a plastic or disposable paper hood that fits over the person's head and is "washed" by a stream of HEPA-filtered air from a self-contained, battery-powered blower. The combination of an impermeable, disposable suit with a plastic flanged PAPR hood provides an extremely high degree of protection against many eventualities, including small particle aerosols, splashes, and massive blood exposure due to hematemesis [27].

Where then do we fix the level of protection for the care-givers, the community, and the environment? Based on the knowledge of pathogenesis of viral diseases in general, the studies of realistic animal models of VHF, and on more limited data on human patients with VHF, there is a natural progression of virus shedding and risk that begins in the incubation period and progresses through to the last stages of infection. During the incubation and early disease process, there is little or no risk except perhaps from blood. Later, viremia is high and there is the possibility of infection of tissues that may lead to external passage of virus; during this period, there are potential risks from secretions, excretions, and shed blood through direct transmission (including larger droplets), indirect transmission, and possibly aerosols. Specific measures for care of these patients have been published [11, 16]. The following is not intended to provide concrete guidelines for care of VHF patients but rather, to suggest a framework for approaching the problem.

### Initial evaluation of the patient

A febrile illness coupled with a history of travel to a VHF-endemic area during the 3 weeks preceding onset should alert the clinician to the possibility of an exotic infectious disease. Identifying an exposure to arthropods or rodents may help in the diagnosis, particularly if there is a history of tick bite or a known rodent reservoir of a pathogenic arenavirus or hantavirus is unusually abundant. Local patterns of virus transmission are important, including detailed geographic and ecologic features, seasonality, or ongoing epidemic transmission. A complete history and physical may suggest other diagnoses. The important, common, potentially lethal diseases to rule out include malaria, typhoid, shigellosis, leptospirosis, and rickettsial infections.

At this preliminary assessment the patient can be evaluated with minimal risk to health care providers, although the wearing of mask, gown, and gloves is desirable, as well as placing the patient in a private cubicle or room during the course of the evaluation. Some of the nonspecific symptoms and signs that might lead one to suspect a VHF include prominent myalgia, severe asthenia, dizziness, postural hypotension, blood pressure lower than normal for the patient, enanthem, tachycardia, or petechiae.

The possibility of viremia exists, and thus clinical laboratory studies should be obtained only to focus on the diagnosis of the specific VHF or to exclude likely, treatable alternative diagnoses. Samples for laboratory tests should be obtained with care, placed in double plastic containers, and taken to the laboratory under the control of a responsible person. In the laboratory, the usual blood-borne disease precautions should be followed, with care taken to prevent the generation of aerosols, and the samples should be disposed of by a method such as sterilization by chemicals or autoclaving or, if freezing for future analysis is required, be placed in clearly labeled, plastic-bagged containers. Helpful but nonspecific findings include proteinuria, thrombocytopenia, hemoconcentration, hypoalbuminemia, elevated AST or leucopenia (although normal or increased white count is a feature of some VHFs). Important tests for other agents include blood cultures, stool cultures, and a malaria smear. Results of the malaria smear must be interpreted with caution, since the presence of parasites does not establish that malaria is the cause of disease in partially immune patients nor does the failure of an inexperienced laboratory to report parasites exclude malaria in non-immune patients.

It is possible to diagnose the VHFs listed in Table 1 (as well as currently unknown but genetically and/or serologically related agents) with antigen-detection enzyme-linked immunosorbent assay (ELISA), immunohistochemistry, IgM capture ELISA, and/or reverse transcription-polymerase chain reaction with established primers [47, 86]. Testing can be completed in 4–24 h in most cases. The more time-consuming conventional virus isolation and paired serologic testing are also useful for diagnosis, recovering virus strains, and confirming unusual findings. The necessary reagents, training, and laboratory facilities for rapid tests are in short supply worldwide.

## Transport

Transporting patients poses specific problems which, like all treatment decisions, are dependent on the specific situation. For example, during the Korean conflict rapid medical evacuation of patients with hemorrhagic fever with renal syndrome by US military helicopters led to early therapy and spared the patients the trauma of ground transportation over rugged roads; in the 1980s, aeromedical evacuation flights from Korea to sites as far distant as Hawaii led to increased mortality [8]. The balance between evacuation to optimum medical facilities and patient deterioration en route is even more delicate in the management of HPS, with its rapid unpredictable decline of clinical status. In the non-hantaviral VHFs, for which disease transmission must also be considered, the goal would be to transfer patients very early in their course of illness when the risk to the patient and to others is minimal. Critically ill patients should preferably not be transported due to the increased mortality associated with moving patients with severely compromised vascular beds. It is also possible to isolate the patient within a device such as a Vickers Medical Stretcher Isolator Unit, during transport but this approach requires anticipatory development and rehearsals for efficient execution.

## Treatment of the uncomplicated patient

Education of the traveling public and medical professionals should lead to most patients being suspected or diagnosed early in the course of their illness. These patients should also be at relatively low risk to transmit disease to health care providers, although the risk of transmission increases as the patient becomes increasingly ill. The patient should be admitted to a private room. An anteroom to serve as a staging area for materials to enter and exit the patient's room and as a place for the staff to don and remove protective gear is a very useful adjunct. Although a room with negative air pressure is not needed at this time it should be considered, if available, to circumvent any subsequent need to transfer the patient to such a room. Gowns, gloves, and mask should be used plus eye protection to prevent skin and mucous membrane exposure to potentially infectious bodily secretions. The level of protection should then be upgraded to anticipate the patient's clinical condition (e.g., increasing respiratory protection in the event of profuse bleeding or respiratory disease). Given the unpredictable nature of these illnesses and the possible progression of infectivity of any given patient, it may be prudent to implement these upgraded measures as soon as feasible. Members of the patient's direct care team should be limited to a small number of selected, trained individuals with a goal of eliminating any parenteral exposure. The danger of a parenteral exposure must be constantly considered. These accidents have been associated with a high probability of transmission, short incubation periods, and severe disease in several episodes. Effective communications with the infection control practitioners and all other health care providers who have contact with either the patient or his body fluids are crucial.

The cordon of protection cannot be limited to direct care providers. All laboratory specimens should be considered infectious and they must be tracked from the point of collection to disposal. These samples should be double-bagged by the person who collected the sample and hand-carried directly to the laboratory for testing. Testing should be limited to assays necessary for patient management. Specimen processing and handling should be done in a manner that minimizes the possibility of larger droplet or small particle aerosol generation. If possible, all specimen handling should occur under a Class 2 laminar flow hood and centrifugation should be performed in covered buckets with O-ring seals. Effluent from automated analyzers should be appropriately decontaminated. Decreasing the viral concentration by using Triton-X-100, beta propionolactone, acidification, heating, or high dose gamma irradiation prior to laboratory tests that are not affected by the chosen method are desirable approaches [2, 22, 48, 57]. Residual clinical samples should be properly stored in a frozen state for future analysis or sterilized by autoclaving, or chemical-inactivating. Similarly, all fomites that are soiled by patient contact or secretions need to be chemically inactivated or autoclaved prior to reuse or disposal. Virus isolation should be attempted by experienced laboratory workers in BSL-4 facilities.

### The severely ill patient

Patients with severe illness pose a more difficult problem. Virus is expected to have potentially infected many organs, perhaps including those critical for virus dissemination, such as the salivary glands, kidney, bladder, sweat glands, and lungs. Hemorrhage may be a feature of their clinical course and viremia is intense in the later stages of most severely ill VHF patients. Although two imported VHF patients have received intensive care management without any precautions beyond the wearing of mask, gowns, and gloves without evidence of secondary infections, the close exposure to body fluids and the many opportunities for generation of larger droplets and small particle aerosols suggest the need for additional precautions as outlined above. The level of protection chosen for this stage, however, must be balanced by practical considerations such as the health care provider's ability to function safely (vis a vis parenteral exposure) without imposing undue physical impediments to provide adequate care.

### Virology of convalescent patients

VHF viruses may remain sequestered in protected immunologic sites, such as the uveal and seminal tracts of convalescent patients; therefore, these patients need appropriate precautions after hospitalization. Lassa virus has been isolated in urine 32 days after disease onset [9]. There is also documented sexual transmission of Junin virus, the causative agent of Argentine HF, to spouses of recovered patients 7–22 days after illness onset [5]. Marburg virus has been isolated from the semen of one patient 2 months [77] following recovery and from another patient 83 days following disease onset [50] from a spouse implicated in a sexually transmitted case [51]. Ebola virus has been isolated 61 days but not 74 days after onset of illness in an infected laboratory worker [21]. Marburg virus has also been isolated from the anterior chamber of the eye from a convalescent Marburg patient with uveitis 80 days after the acute disease began [28].

There is no evidence for long-term asymptomatic viral shedding, such as that seen among persons with chronic hepatitis B virus infection; therefore, precautions to prevent sexual transmission can be safely terminated when virus is no longer detected in genital secretions, probably within 2–3 months. Viral isolation from breast milk or reports of breast milk acquired Junin and Lassa infection among nursing infants [49, 60] suggest that, if possible, breast-feeding should be interrupted until the breast milk is determined to be sterile. Moreover, physicians such as ophthalmologists should recognize that post-infectious sequelae in these patients may be due to the presence of residual virus.

## Protection of the environment

Since VHF are exotic, the risk of establishing a natural zoonotic transmission cycle must always be considered. Each virus has an established geographic range which implies that there are constraints on spread of the virus, and these are imposed by the availability of suitable reservoir/vector species. Thus, the probability of a distant introduction must be judged to be low in general. There are, however, several exceptions that should be considered. First, yellow fever virus was introduced into the Americas from Africa with its urban vector *Aedes aegypti* and is now also successfully established in the jungles of South America and vectored by native mosquitoes. Second, among the arenaviruses, lymphocytic choriomeningitis virus from the house mouse has been able to persist in Syrian hamster colonies, and Lassa virus from African rodents can chronically infect selected genotypes of laboratory mice [67]. Finally, because the reservoir(s) and natural history of filoviruses remain unknown, it is impossible to even speculate as to their risk. The prudent course is to minimize any possibility of an introduction to a potentially susceptible venue.

All bodily excretions from a patient with VHF should be considered infectious and should be inactivated prior to disposal to 1) prevent subsequent accidental infections (e.g., undessicated blood from acutely infected Ebola patients may be infectious for up to 1 month at ambient temperature [R. Swanepoel, unpubl.data]) and 2) prevent establishment of disease within a local susceptible vector [13]. Patients with RVF as well as yellow fever or dengue should be screened from mosquitoes to prevent further arthropod transmission. All effluent should be disinfected prior to disposal into a municipal sewer system or septic tank by adding disinfectant prior to use [84] or using chemical toilets. This level of precaution should continue for 6 weeks of convalescence or until the patient is virologically negative [21].

## References

1. Bannister BA (1993) Stringent precautions are advisable when caring for patients with viral haemorrhagic fevers. Rev Med Virol 3: 3–6
2. Barry M, Russi M, Armstrong L, Geller D, Tesh R, Dembry L, Gonzalez JP, Khan AS, Peters CJ (1995) Brief report: treatment of a laboratory-acquired Sabiá virus infection. N Engl J Med 333: 294–296
3. Bazhutin NB, Belanov EF, Spiridonov VA, Voltenko AV, Krivenchuk NA, Krotov SA, Omel'chenko NI, Teresbchenko AY, Khomichev VV (1992) The influence of the methods of experimental infection with Marburg virus on the course of illness in green monkeys. Vopr Virus 3: 153–156
4. Berendt RF, Brown JL, Jemski JV (1981) Assessment of airborne microbial agents of potential BW threat. 1981 USAMRIID Annual Report. United States Army Medical Research Institute for Infectious Diseases, Federick, Maryland, pp 19–22
5. Briggiler A, Enria D, Feuillade MR, Maiztegui J (1987) Contagio interhumano e infección clinica con virus Junín (VJ) en matrimonios residentes en el área endemica de fiebre hemorrágica argentina (FHA). Medicina 47: 565
6. Brown JL (1981) Characteristics of aerosol induced Rift Valley fever infections. 1980 USAMRIID Annual Report. United States Army Medical Research Institute for Infectious Diseases, Federick, Maryland, pp 223–230
7. Brown JL, Dominik JW, Morrissey RL (1981) Respiratory infectivity of a recently isolated Egyptian strain of Rift Valley fever virus. Infect Immun 33: 848–853
8. Bruno P, Hassell LH, Brown J, Tanner W, Lau A (1990) The protean manifestations of haemorrhagic fever with renal syndrome. A retrospective review of 26 cases from Korea. Ann Int Med 113: 385–391
9. Buckley SM, Casals J (1970) Lassa fever, a new virus disease of man from west Africa. III. Isolation and characterization of the virus. Am J Trop Med Hyg 19: 680–691
10. Carey DE, Kemp GE, White HA, Pinneo L, Addy RF, Fom ALMD, Stroh G, Casals J, Henderson BE (1972) Lassa fever. Epidemiologic aspects of the 1970 epidemic, Jos, Nigeria. Trans R Soc Trop Med Hyg 66: 402–408
11. CDC (1988) Management of patients with suspected viral haemorrhagic fever. MMWR 37(no. S-3): 1–15
12. CDC (1990) Update: filovirus infection in animal handlers. MMWR 39: 221
13. CDC (1990) Mosquito-transmitted malaria – California and Florida, 1990. MMWR 40: 106–108
14. CDC (1995) Outbreak of Ebola viral haemorrhagic fever – Zaire, 1995. MMWR 44: 381–382
15. CDC (1995) Update: outbreak of Ebola viral haemorrhagic fever – Zaire, 1995. MMWR 44: 468–469, 475
16. CDC (1995) Update: management of patients with suspected viral haemorrhagic fever – United States. MMWR 44: 475–479
17. Couch RB, Cate TR, Douglas RG, Gerone PJ, Knight V (1966) Effect of route of inoculation on experimental respiratory viral disease in volunteers and evidence for airborne transmission. Bacteriol Rev 30: 517–529

18. Cummins D, Bennett D, Fisher-Hoch SP, Farrar B, Machin SJ, McCormick JB (1992) Lassa fever encephalopathy: clinical and laboratory findings. J Trop Med Hyg 95: 197–201

19. Dalgard DW, Hardy RJ, Pearson SL, Pucak GJ, Quander RV, Zack PM, Peters CJ, Jahrling PB (1992) Combined simian haemorrhagic fever and Ebola virus infection in cynomolgus monkeys. Lab Anim Sci 42: 152–157

20. Douglas RG Jr, Wiebenga NH, Couch RB (1965) Bolivian haemorrhagic fever probably transmitted by personal contact. Am J Epidemiol 82: 85–91

21. Emond RTS, Evans B, Bowen ET, Lloyd G (1977) A case of Ebola virus infection. Br Med 2: 541–544

22. Elliott LH, McCormick JB, Johnson KM (1982) Inactivation of Lassa, Marburg, and Ebola viruses by gamma irradiation. J Clin Microbiol 16: 704–708

23. Fenner F (1948) The clinical features of mouse-pox (infectious ectromelia of mice) and the pathogenesis of the disease. J Pathol Bacteriol 60: 529–552

24. Fisher-Hoch SP (1993) Stringent precautions are not advisable when caring for patients with viral haemorrhagic fevers. Rev Med Virol 3: 7–13

25. Fisher-Hoch SP, Brammer TL, Trappier SG, Hutwagner LC, Farrar BB, Ruo SL, Brown BG, Hermann LM, Perez-Oronoz GI, Goldsmith CS, Hanes MA, McCormick JB (1992) Pathogenic potential of filoviruses: role of geographic origin of primate host and virus strain. J Infect Dis 166: 753–763

26. Fisher-Hoch SP, Platt GS, Neild GH, Southee T, Baskerville A, Raymond RT, Lloyd G, Simpson DIH (1985) Pathology of shock and hemorrhage in a fulminating viral infection (Ebola). J Infect Dis 152: 887–894

27. Foberg U, Fryden A, Isaksson B, Jahrling P, Johnson A, McKee K, Niklasson B, Norman B, Peters CJ, Bengtsson M (1991) Viral haemorrhagic fever in Sweden; experience from management of a case. Scand J Infect Dis 23: 143–151

28. Gear JSS, Cassel GA, Gear AJ, Trappler B, Clausen L, Meyers AM, Kew MC, Bothwell TH, Sher R, Miller GB, Schneider J, Koornhof HJ, Gomperts ED, Isaäcson M, Gear JHS (1975) Outbreak of Marburg virus disease in Johannesburg. Br Med 4: 489–493

29. Geisbert TW, Jahrling PB, Hanes MA, Zack PM (1992) Association of Ebola-related Reston virus particles and antigen with tissue lesions of monkeys imported to the United States. J Comp Pathol 106: 137–152

30. Gerone PJ, Couch RB, Keefer GV, Douglas RG, Derrenbacher EB, Knight V (1966) Assessment of experimental and natural viral aerosols. Bacteriol Rev 30: 576–584

31. González LE, Sabattini MS, Fain Binda YJC (1967) Recuperación de virus Junín de pacientes de fiebre hemorrágica Argentina con confirmación serológica. Rev Soc Argent Biol 43: 261–270

32. Helmick CG, Webb PA, Scribner CL, Krebs JW, McCormick JB (1986) No evidence for increased risk of Lassa fever infection in hospital staff. Lancet 8517: 1202–1205

33. International Commission (1978) Ebola haemorrhagic fever in Zaire, 1976. Bull World Health Organ 56: 271–293

34. Jaax N, Davis K, Geisbert T, Vogel A, Jaax G, Topper M, Jahrling P (1996) Experimental infection of rhesus monkeys by the oral and conjunctival route of exposure. Arch Pathol Lab Med 120: 140–155

35. Jaax M, Geisbert T, Jahrling P, Geisbert J, Steele K, McKee K, Negley D, Johnson E, Peters C (1995) Natural transmission of Ebola virus (Zaire strain) to monkeys in a biocontainment laboratory. Lancet 346: 1669–1671

36. Jahrling PB, Geisbert TW, Jaax NK, Hanes MA, Ksiazek TG, Peters CJ (1996) Experimental infection of cynomolgus macaques with Ebola-Reston filoviruses from the

1989–1990 U.S. epizootic. In: Schwarz TF, Siegl G (eds) Imported virus infections. Springer, Wien New York, pp 115–134 (Arch Virol [Suppl] 11)

37. Jahrling PB, Hesse RA, Eddy GA, Johnson KM, Calls RT, Stephen EL (1980) Lassa virus infection of rhesus monkeys: pathogenesis and treatment with Ribavirin. J Infect Dis 141: 580–589

38. Jewett DL, Heinsohn, Bennett C, Rosen A, Neuilly (1992) Blood containing aerosols generated by surgical techniques; a possible infectious hazard. Am Ind Hyg Assoc J 53: 228–231

39. Johnson E, Jaax N, White J, Jahrling P (1995) Lethal experimental infections of rhesus monkeys by aerosolized Ebola virus. Int J Exp Pathol 76: 227–236

40. Johnson KM, McCormick JB, Webb PA, Smith ES, Elliott LH, King IJ (1987) Clinical virology of Lassa fever in hospitalized patients. J Infect Dis 155: 456–463

41. Keefer GV, Zebarth GL, Allen WP (1972) Susceptibility of dogs and cats to Rift Valley fever by inhalation or ingestion of virus. J Infect Dis 125: 307–309

42. Kenyon RH, McKee KT Jr, Zack PM, Rippy MK, Vogel AP, York C, Meegan J, Crabbs C, Peters CJ (1992) Aerosol infection of rhesus macaques with Junin virus. Intervirology 33: 23–31

43. Knight V (1973) Airborne transmission and pulmonary deposition of respiratory viruses. In: Knight V (ed) Viral and mycoplasma infection of the respiratory tract. Lea and Febiger, Philadelphia, pp 1–9

44. Ksiazek TG, Peters CJ, Rollin PE, Zaki S, Nichol ST, Spiropoulou CF, Morzunov S, Feldmann H, Sanchez A, Khan AS, Mahy BWJ, Wachsmuth K, Butler J (1995) Identification of a new north American hantavirus that cases acute pulmonary insufficiency. Am J Trop Med Hyg 52: 117–123

45. Kuehne RW (1973) A biological containment facility for studying infectious disease. Appl Microbiol 26: 239–243

46. Langmuir AD (1980) Changing concepts of airborne infection of acute contagious diseases: a reconsideration of classic epidemiologic theories. Ann NY Acad Sci 353: 35–44

47. Lennette EH (ed) (1992) Laboratory diagnostics of viral infections, 2nd ed. Marcel Dekker, New York

48. Lloyd G, Bowen ETW, Slade JHR (1982) Physical and chemical methods of inactivating Lassa virus. Lancet 8280: 1046–1048

49. Maiztegui JI, Voeffrey JR, Fernández NJ, Barrera Oro JG (1973) Aislamiento de virus Junín a partir de leche materna. Medicina 33: 659–660

50. Martini GA, Schmidt H (1968) Speratogene Übertragung des Marburg Virus. Klin Wochenschr 46: 391

51. Martini GA (1971) Marburg virus disease. Clinical syndrome. In: Martini GA, Siegart R (eds) Marburg virus disease. Springer, Berlin Heidelberg New York, pp 1–9

52. McKee KT Jr, Mahlandt BG, Maiztegui JI, Eddy GA, Peters CJ (1985) Experimental Argentine haemorrhagic fever in Rhesus macaques: viral strain dependent clinical response. J Infect Dis 152: 218–221

53. McKee KT Jr, Mahlandt BG, Maiztegui JI, Green DE, Peters CJ (1987) Virus-specific factors in experimental Argentine haemorrhagic fever in Rhesus macaques. J Med Virol 22: 99–111

54. McKinney RW, Barkley WE, Wedum AG (1991) The hazards of infectious agents in microbiological laboratories. In: Block SS (ed) Disinfections, sterilization, and preservation, 4th ed. Philadelphia, Lea and Febiger, pp 748–756

55. Meegan JM (1979) The Rift Valley fever epizootic in Egypt 1977–78. 1. Description of the epizootic and virological studies. Trans R Soc Trop Med Hyg 73: 618–623

56. Miller WS, Demchak P, Rosenberger CR, Dominink JW, Bradshaw JL (1963) Stability and infectivity of airborne yellow fever and rift valley fever viruses. Am J Hyg 77: 114–121

57. Mitchell SW, McCormick JB (1984) Physicochemical inactivation of Lassa, Ebola, and Marburg viruses and effect on clinical laboratory analyses. J Clin Microbiol 20: 486–489

58. Monath TP (1975) Lassa fever: review of epidemiology and epizootiology. Bull World Health Organ 52: 577–592

59. Monath TP, Maher M, Casals J, Kissling RE, Cacciapuoti A (1974) Lassa fever in the Eastern Province of Sierra Leone, 1970–1972. II. Clinical observations and virologic studies on selected hospital cases. Am J Trop Med Hyg 23: 1140–1149

60. Monson MH, Cole AK, Frame JD, Serwint JR, Alexander S, Jahrling PB (1987) Pediatric Lassa fever: a review of 33 liberian cases. Am J Trop Med Hyg 36: 408–415

61. Morrow PE (1980) Physics of airborne particles and their deposition in the lung. Ann NY Acad Sci 353: 71–79

62. Muyembe T, Kipasa M, The International Scientific and Technical Committee, WHO Collaborating Centre for Haemorrhagic Fevers (1995) Ebola haemorrhagic fever in Kikwit, Zaire [letter]. Lancet 345: 1448

63. Nuzum EO, Rossi CA, Stephenson EH, LeDuc JW (1988) Aerosol transmission of Hantaan and related viruses to laboratory rats. Am J Trop Med Hyg 38: 636–640

64. Peters CJ (1994) Arenaviruses. In: Osterhaus ADME (ed) Virus infections of rodents and lagomorphs. Elsevier, Amsterdam, pp 321–341 (Virus Infections of Vertebrates, vol 5)

65. Peters CJ, Jones D, Trotter R, Donaldson J, White J, Stephen F, Slone TW Jr (1988) Experimental Rift Valley fever in Rhesus macaques. Arch Virol 99: 31–44

66. Peters CJ, Johnson ED, Jahrling PB, Ksiazek TG, Rollin PE, White J, Hall W, Trotter R, Jaax N (1993) Filoviruses. In: Morse S (ed) Emerging viruses. Oxford University Press, New York, pp 159–175

67. Peters CJ, Johnson ED, McKee KT Jr (1991) Filoviruses and management of viral haemorrhagic fevers. In: Belshe R (ed) Textbook of human virology, 2nd ed. Mosby Year Book, St. Louis, pp 699–712

68. Peters CJ, Kuehne RW, Mercado RR, Le Bow RH, Spertzel RO, Webb PA (1974) Haemorrhagic fever in Cochabamba Bolivia, 1971. Am J Epiz 99: 425–433

69. Peters CJ, LeDuc JW (1996) Viral haemorrhagic fevers: persistent problems, persistence in reservoirs. In: Mahy BWJ, Compans RW (eds) Immunobiology and pathogenesis of persistent virus infections. Harwood, Chur (in press)

70. Petersen NJ (1980) An assessment of the airborne route in hepatitis B transmission. Ann NY Acad Sci 353: 157–166

71. Pokhodynaev VA, Gonchar NI, Pshenichnov VA (1991) Experimental study of Marburg virus contact transmission. Vopr Virus 6: 506–508

72. Richmond JY, McKinney (Eds) (1993) Biosafety in microbiological and biomedical laboratories, 3rd ed. HHS publication number (CDC) 8393–8395. US Government Printing Office, Washington

73. Shepard AJ, Swanepoel R, Shepard SP, Leman PA, Blackburn NK, Hallett AF (1985) A nosocomial outbreak of Crimean-Congo haemorrhagic fever at Tygerberg hospital. Part V. Virological and serological observations. S Afr Med J 68: 733–736

74. Simpson DIH (1969) Marburg agent disease: in monkeys. Trans R Soc Trop Med Hyg 63: 303–309

75. Simpson DIH (1970) Marburg virus: a review of laboratory studies. In: Balner H, Beveridge WIB (eds) Infections and immunosuppression in subhuman primates. Scandinavian University Press, Munksgaard, pp 39–44

76. Simpson DIH, Zlotnik I, Rutter DA (1968) Vervet monkey disease. Experimental infection of guinea-pigs and monkeys with the causative agent. Br Exp Pathol 49: 458–464

77. Smith DH, Johnson BK, Isaacson M, Swanepoel R, Johnson KM, Killey M, Bagshawe A, Siongok T, Keruga WK (1982) Marburg-virus disease in Kenya. Lancet 8276: 816–820

78. Stephenson EW (1982) Assessment of airborne microbial agents of potential BW threat. 1982 USAMRIID Annual Report. United States Army Medical Research Institute for Infectious Diseases, Federick, Maryland, pp 77–80

79. Stephenson EH, Larson EW, Dominik JW (1984) Effect of environmental factors on aerosol-induced Lassa virus infection. J Med Virol 14: 295–303

80. Stojković LJ, Bordkoški M, Gligić A, Stefanović Ž (1971) Two cases of Cercopithecus-monkeys-associated haemorrhagic fever. In: Martini GA, Siegart R (eds) Marburg virus disease. Springer, Berlin Heidelberg New York, pp 24–33

81. Suleiman MN, Muscat-Barron JM, Harries JR, Satti AG, Platt GS, Bowen ET, Simpson DI (1980) Congo-Crimean haemorrhagic fever in Dubai. An outbreak at the Rashid Hospital. Lancet 8201: 939–941

82. Swanepoel R, Shepard AJ, Leman PA, Shepard SP, McGillivrary GM, Erasmus MJ, Searle CA, Gill DE (1987) Epidemiologic and clinical features of Crimean-Congo haemorrhagic fever in south Africa. Am J Trop Med Hyg 6: 120–132

83. Van Eeden PJ, Joubert JR, va de Wal BV, King JB, de Kock A, Groenewald JH (1985) A nosocomial outbreak of Crimean-Congo haemorrhagic fever at Tygerberg Hospital. Part 1. Clinical features. S Afr J Med 68: 711–717

84. Wallis C, Melnick JL, Rao VC, Sox TE (1985) Method for detecting viruses in aerosols. Appl Environ Microbiol 50: 1181–1186

85. WHO/International Study Team (1978) Ebola hemorrhagic fever in Sudan, 1976. Bull World Health Organ 56: 247–270

86. Zaki S, Marty AM (1995) New Technology for the diagnosis of infectious diseases. In: Doerr D, Seifert G (eds) Tropical pathology. Springer, Berlin Heidelberg New York, pp 127–154

Authors' address: Dr. C. J. Peters, Centers for Disease Control and Prevention, Mailstop A-26, 1600 Clifton Road, Atlanta, GA 30333, U.S.A.

# Hepatitis viruses and human immunodeficiency virus

Arch Virol (1996) [Suppl] 11: 171–179

# Relative importance of the enterically transmitted human hepatitis viruses type A and E as a cause of foreign travel associated hepatitis

### B. C. A. Langer and G. G. Frösner

Max von Pettenkofer Institute for Hygiene and Medical Microbiology,
Ludwig Maximilians University, Munich, Federal Republic of Germany

**Summary.** Hepatitis contracted during a stay abroad may be caused by a wide range of pathogens including viruses, bacteria, protozoa or helminths. In many cases, the etiological agent primarily infects other target organs and tissues, involving the liver either as part of a disseminated infection or secondarily to mechanical biliary tract obstruction. The article focuses on enterically transmitted hepatitis caused by the primarily hepatotropic human hepatitis viruses type A and E and discusses their importance in travel-related disease.

## Hepatitis A

### Epidemiology

Viral hepatitis type A occurs in all parts of the world, yet the epidemiologic features vary in different geographic regions, reflecting determinants such as socioeconomic conditions and geographic factors. In countries, where living conditions are crowded and environmental sanitation is generally poor, most hepatitis A virus (HAV) infections occur at an early age, when asymptomatic courses are common, and close to 100% of children acquire protective immunity during the first decade of life. In these areas, distinct outbreaks are rare, and disease related to hepatitis A is uncommon. As hygienic and sanitary standards improve, exposure and thus infection shift to older age groups with increasing incidence of clinically apparent disease. In most industrialized countries, improvements in socioeconomic standard over the recent years have been associated with a marked decline in prevalence and incidence of hepatitis A, leaving a high proportion of the population susceptible to infection. Related to lower socioeconomic conditions in the past, 60–80% of the people born before 1950 show evidence of prior HAV infection, whereas 80–90% of the less than 40-year-olds are seronegative and thus susceptible [7, 16]. In this low prevalence setting, infections tend to occur among specific risk groups, notably travellers, intravenous drug users or homosexuals. In some European countries, a history of recent travel to hepatitis A endemicity areas accounts for up to two thirds of reported cases, compared to less than 5% in the USA [7, 16]. The risk for unprotected travellers has been estimated to vary between 0.3–0.6% and 2.0% per month of stay in a developing country, the lower figure applying to tourists and business persons staying in good and even luxury hotels, the higher to trampers staying at places with poor hygienic conditions [7, 18]. Thus, after diarrhoea and malaria, hepatitis A represents the third most important foreign travel associated infection. Considering the clinical impact, it is assumed that 90% of all HAV infections among adult travellers are symptomatic. The percentages of relatively mild, moderate, severe, i.e. requiring hospitalization, and fulminant courses are estimated at 50%, 30%, 19.5% and 0.5%, respectively.

Another route of HAV "importation" involves children of foreign workers or immigrants from hepatitis A endemic countries who contracted the infection abroad. Exposure to these "imported cases", e.g. in day care centers, may lead to considerable secondary transmission. Because infected young children are usually asymptomatic, outbreaks are often only recognized when older playmates, parents or employees become clinically ill.

The principal mode of spread for hepatitis A is by the faecal-oral route, most commonly by person-to-person contact, especially under crowded and poorly hygienic living conditions. Faeces of infected persons can contain copious amounts of infectious virions (up to $10^8$/ml), with the highest titers excreted during the late incubation period, i.e. prior to any clinical, biochemical or serological evidence of infection. Communicability closely correlates with this period of maximal virucopria, subsiding substantially with the onset of disease. Infections due to consumption of faecally contaminated food or beverages are also common. The vehicles of transmission are most often uncooked or inadequately heated foods, and foods contaminated after cooking by infected handlers in the incubation stage of their disease. Waterborne infections have been reported, associated with the use of faecally contaminated drinking water, as well as with swimming in contaminated swimming pools or sewage-polluted lakes [2, 11]. Transmission of HAV by externally contaminated coprophagous insects remains unconfirmed.

*Virology*

Hepatitis A virus has recently been classified as the prototype virus of the new genus hepatovirus within the family Picornaviridae.

The virion is a non-enveloped, spherical particle with a diameter of about 27 nm. The single-stranded, positive-sense RNA genome of about 7.5 kb contains one large open reading frame encoding a polyprotein precursor that is co-translationally processed into the capsid proteins at the N-terminal end, and the non-structural proteins at the C-terminus. Assembled particles display an immunodominant, conformationally dependent neutralization epitope. Based on genetic relatedness, HAV isolates have been assigned to seven unique genotypes distinguished by more than 15% nucleotide sequence diversity.

Within each of these genotypes, subtypes differing in approximately 7.5% of base positions can be defined. Virus strains within these subtypes exhibit less than 3% diversity over periods of up to 15 years [15, 21]. Four of the seven HAV genotypes have been recovered from infected humans, the remaining three have been isolated from old world monkeys. Molecular-epidemiological studies found most of the HAV isolates in the USA and China belonging to genotype I, whereas multiple genotypes have been identified in Europe and Japan, probably representing viruses "imported" from more endemic regions [15].

However, hepatitis A virus exhibits a remarkable antigenic stability. Only a single serotype has been recognized among human strains isolated from widely separated geographic regions. Immune serum globulin preparations produced in the United States have been shown to protect against hepatitis A anywhere in the world. Moreover, cross-neutralization studies using geographically distinct isolates confirmed the close antigenic relatedness.

Another notable feature of the HAV particle is its exceptional physico-chemical stability. The virus has been shown to survive on surfaces for up to one month in a dried state under ambient conditions. In water, sewage, soils, marine sediment or live shellfish, it may retain infectivity for 12 weeks to 10 months [6, 12]. HAV is also relatively resistant to heat or chemical inactivation. It is stable at 60 °C for 1 h, and is only partially inactivated after 10–12 h. Variation in pH between 3 and 10, or even brief exposure to pH 1, have no demonstrable effect on structural integrity or biological activity of the virus.

### Clinical manifestation of hepatitis A infection

Hepatitis A is typically a relatively benign, self-limited disease with a case fatality rate of less than 0.1%. In general, the frequency and severity of clinical manifestation progressively increase with the age at infection. In the very young, infections are usually asymptomatic, whereas the rare fulminant and fatal courses occur almost exclusively in those infected above the age of 50.

After an incubation period of 15–50 days, the illness begins relatively abrupt with non-specific symptoms such as fatigue and malaise, fever, anorexia, abdominal pain, nausea and vomiting. These prodromi are followed by more specific signs of dark urine and clay-coloured stool. Jaundice of the sclera and skin may be noted when the serum bilirubin exceeds 2.5–3.0 mg/dl. Just prior to the icteric phase, serum aminotransferase activities become elevated.

Almost all patients recover within 1–2 months, although in some the convalescence phase may be prolonged for several months ("post-hepatitis syndrome"). Yet there is no evidence of viral persistence or progression to chronic liver disease.

Some patients experience a relapsing course with recurrent clinical symptoms, renewed elevation of liver enzymes and virus shedding occuring 1–3 months after the initial presentation. The relapse may last for several months, but complete recovery eventually ensues.

Occasionally, the infection presents with a cholestatic picture and jaundice for up to 4 months, but the condition resolves completely without permanent sequelae.

A single infection appears to generate lifetime protection against clinical disease.

### Diagnosis

A specific diagnosis of acute hepatitis A is most easily and economically made by testing for virus-specific IgM antibody. This acute-phase marker is almost invariably present in serum at the onset of clinical illness, declining to nondetectable levels within 3–6 months in most patients. Very rarely, IgM may persist at low or borderline levels for up to 18 months.

IgG class anti-HAV can also be detected very early in the course of the infection, and generally persists for life providing protective immunity.

Sensitive, solid-phase IgM capture immunoassays or competitive inhibition immunoassays for the detection of total anti-HAV are commercially available.

In addition, hepatitis A virus antigen can be detected in faecal specimens by enzyme-linked or radioimmunoassay. However, since virus excretion is severely curtailed after onset of symptoms, failure to detect antigen does not exclude the diagnosis of acute hepatitis A.

Molecular-virological methods such as HAV RNA detection by reverse transcription-polymerase chain reaction, or virus isolation in cell culture are neither practical nor necessary for routine diagnostic purposes.

### Prevention

The risk of hepatitis A infection can be reduced by taking the same general hygienic precautions as applied to prevent transmission of any enteric infection. Potentially faecally contaminated food or beverages should be avoided. In this regard, the old colonial rule of "boil it, cook it, peel it – or forget it" is still applicable. Careful handwashing practices, especially prior to food preparation or meals, represent another preventive measure, the more so since the virus has been shown to survive intact on fingertips and other surfaces.

Passive immunization by the administration of normal human immunoglobulin has proven to be effective for preventing hepatitis A when given either before or within 10–14 days after exposure to the virus. While providing almost immediate protection, the protective effect is short-lived, requiring repeated injections when prolonged exposure is anticipated. A dose of 0.02 ml/kg body weight will provide protection for up to 3 months, while a dose of 0.06 ml/kg may protect against disease for up to 6 months [22].

With the licence of a formalin-inactivated whole virus vaccine (HAVRIX/SB, Rixensart/Belgium) in several European countries, an effective alternative became available. The basic course of immunization consists of two vaccinations one month apart, with a booster injection 6–12 months later. The time interval between the first two doses can be shortened to 2 weeks in order to achieve immunity more rapidly. Active immunization has been shown to be safe and yielding both, a high seroconversion rate and high antibody levels, while causing no serious adverse reactions. A nearly 100% seroconversion is reached 2–4 weeks after two doses. Protection is thought to last for at least 5–10 years after a complete three-dose-regimen [1]. Following administration of HAVRIX with the double antigen dose (1440 Elisa-units/ml) in a 0–6 months schedule, the immune response develops more rapidly with 77–96% seroconversion two weeks after one dose. An analogous product has been developed by Merck Sharp & Dohme (VAQTA/MSD, West Point/USA).

Two additional hepatitis A vaccines are available so far: an inactivated vaccine formulated in immunostimulating, reconstituted influenza virosomes (EPAXAL/Swiss Serum and Vaccine Institute, Berne) and a live attenuated vaccine developed in China.

Cost-effectiveness analyses have shown that for travellers expected to travel 3 or more times in 10 years (average duration of stay: 3–4 weeks) or for trips exceeding a period of 6 months, active immunization is the most cost-effective preventive option. Passive immunization by the administration of immunoglobulin remains the most cost-effective measure for those expected to travel not more than twice and for stays of less than 3 months duration every 10 years. Travellers born before 1950 or those originating from hepatitis A endemic countries may be screened for anti-HAV prior to immunization in order to save unnecessary vaccinations.

## Hepatitis E

### Epidemiology

Viral hepatitis type E, initially termed enterically transmitted non-A, non-B hepatitis (ET-NANBH), is a major cause of acute hepatitis observed among young to middle-aged adults in developing countries. Endemic regions include the Indian subcontinent with Nepal and Pakistan, Central and South-East Asia, North, West and East Africa, the Middle East and Central America. 7–24% of the population in these areas show serologic evidence of prior exposure [8, 13]. Hepatitis E occurs most frequently in epidemics often involving several thousand cases, with outbreaks observed predominantly during or after the rainy season, when environmental conditions promote spread. In endemic areas, the majority of acute outbreak-associated non-A, non-B hepatitis cases are due to hepatitis E. The reported figures range from 61% in Africa to 92% on the Indian subcontinent [8]. In addition, hepatitis E also accounts for a substantial proportion of acute sporadic hepatitis cases in endemicity regions. 50–75% of acute sporadic non-A, non-B, non-C hepatitis cases have been diagnosed as due to hepatitis E infection.

No outbreak of clinically overt hepatitis E has ever been documented in the USA, Canada, Europe, Japan, Australia and New Zealand. In these countries, only sporadic cases are occasionally observed among immigrants and travellers returning from disease-endemic areas [5, 17, 24]. However, seroprevalence studies suggest that 0.4–2.5% of the population have been exposed to the virus [9, 13, 23, 25]. Recent findings in Germany (Frösner et al., unpubl. res.), the Netherlands [24], Italy [25] and Greece [19] suggest that a lacking foreign travel history does not necessarily seem to exclude the diagnosis of hepatitis E infection. Sporadic cases in Italy and Spain have been attributed to the consumption of shellfish from sewage-polluted waters.

Hepatitis E is transmitted by the faecal-oral route with faecally contaminated drinking water being the primary vehicle of transmission. The infection may also be associated with the ingestion of faecally contaminated food. Person-to-person contact is not believed to be a common mode of spread, as only a very low frequency of clinical disease has been observed in case contacts. However, clinically inapparent secondary transmission may occur, because elevated serum transaminase levels have been detected in household contacts of hepatitis E patients. The highest attack rate of clinically apparent hepatitis E is observed among individuals between 15 and 40 years of age. However, children may also develop icteric disease as well as inapparent or subclinical hepatitis [20].

Another important, but unexplained observations is the unusually high case fatality rate in infected pregnant women. Up to 20% of women with hepatitis E in their third trimester of pregnancy succumb, compared to an overall case fatality rate of 0.5–3%.

## Virology

The aetiological agent, named hepatitis E virus (HEV), has recently been identified and characterized [14]. Yet the taxonomic placement is still uncertain.

The virion is a nonenveloped, spherical particle with a diameter of 27–34 nm. The genome is a single-stranded, positive-sense RNA molecule of approximately 7.5 kb, without obvious sequence similarity to any known virus. Three partially overlapping open reading frames (ORF) have been identified. ORF-1 extending approximately 5 kb from the 5′ end is presumed to encode non-structural proteins. ORF-2 at the 3′ end comprises approximately 2 kb and is thought to encode the major structural protein(s) of HEV. The function of the third open reading frame of only 369 nucleotides is presently unknown. ORF2 and ORF3 have been found to encode immunodominant antigens useful for the diagnosis of hepatitis E.

Cloning and sequencing of HEV isolates from different geographic regions revealed a high degree of sequence divergence between Asian and Mexican isolates. The Mexican strain shows an overall nucleic acid identity of 76% and 77% with the Burma and Pakistan strains, respectively. Comparison of the deduced amino acid sequences reveals 83% and 84% amino acid identity in ORF1, 93% in ORF2 and 87% in ORF3, respectively.

Yet, HEV isolates from geographically distinct regions of the world have been shown to possess at least one major cross-reactive epitope.

Hepatitis E virus is excreted in faeces. However, limited information is available about the pattern of virus excretion. Virus-like particles were most often detected in stool specimens collected between day 4 prior to and day 6 after the onset of clinical symptoms. The virus appears to be extremely labile to unfavourable environmental conditions, including the enzymatic milieu in the digestive tract. So, the virus concentrations in stool samples are rather low, and particles are often shed in a degraded form and intermittent pattern.

## Clinical manifestation of HEV infection

After an incubation period of approximately 5–6 weeks, an acute icteric hepatitis, similar to hepatitis A, develops. Cholestasis seems to be more prominent in hepatitis E. The serum bilirubin may exceed 20 mg/dl. The aminotransferase levels are often only modestly elevated. The course is usually uncomplicated and self-limiting with clinical and biochemical recovery. Chronicity or viral persistence do not develop. However, a high rate of fulminant disease occurs in infected pregnant women, especially in the third trimester of pregnancy. These women show a high incidence of disseminated intravascular coagulation with bleeding diathesis. The complete spectrum of HEV-induced illness is not yet known, particularly with regard to subclinical infections.

## Diagnosis

Cloning and sequencing of the entire HEV genome allowed development and application of serodiagnostic assays for the detection of HEV-specific antibodies.

Test formats such as enzyme immunoassay or Western Blot using ORF2 and ORF3 derived recombinant proteins or synthetic peptides have been shown to detect anti-HEV antibodies in acute- and convalescent-phase sera collected from outbreaks worldwide and also in acute sporadic non-A, non-B hepatitis cases. IgG anti-HEV antibodies remained detectable for up to 2 years, in one case for up to 4 years, after the acute phase of illness. IgM antibodies were detected in about 75% of patients within 1 month of onset of jaundice, in 50% of cases after 2 months and in 6% after 6 months [3]. 8 months after onset of jaundice, IgM was no longer detectable.

At present, it is not known whether and to what extent seropositivity reflects immunity against HEV infection.

HEV RNA can be detected in acute phase faecal and serum samples by polymerase chain reaction. Recently, we found HEV RNA for up to 2 weeks after onset of clinical symptoms in stool and serum samples of a patient with acute hepatitis E. Sequencing of the PCR product may allow to trace foci of infection.

## Prevention

At present, the only possible prophylactic measures for travellers to hepatitis E endemic regions are good personal hygiene, boiling or disinfection of drinking water and avoidance of foods which cannot be peeled or which are not cooked.

It is not known whether hepatitis E, like hepatitis A, might be modified or prevented by administration of immune serum globulin. Certainly, immune serum globulin prepared within non-endemic regions is unlikely to contain significant levels of anti-HEV antibodies, because only a very small proportion of the population turns out to be seropositive. On the other hand, it is not clear whether immune serum globulin would be protective if prepared from plasma collected in countries where HEV infection is more prevalent. Candidate vaccines for active immunization have not yet been described.

## References

1. Andre FE, D'Hondt E, Delem A, Safary A (1992) Clinical assessment of safety and efficacy of an inactivated hepatitis A vaccine: rationale and summary of findings. Vaccine 10 [Suppl 1]: 160–168
2. Bryan JA, Lehmann JD, Setiady IF, Hatch MH (1974) An outbreak of hepatitis A associated with recreational lake water. Am J Epidemiol 99: 145–154
3. Centers for Disease Control (1987) Enterically transmitted non-A, non-B hepatitis – East Africa. Morb Mort Wkly Rep 36: 241–244
4. Centers for Disease Control (1987) Enterically transmitted non-A, non-B hepatitis – Mexico. Morb Mort Wkly Rep 36: 597–602
5. Centers for Disease Control (1993) Hepatitis E among US travellers 1989–1992. Morb Mort Wkly Rep 42: 1–4
6. Enriquez R, Frösner GG, Hochstein-Mintzel V, Riedemann S, Reinhardt G (1992) Accumulation and persistence of hepatitis A virus in mussels. J Med Virol 37: 174–179
7. Frösner GG, Roggendorf M, Frösner HR, Gerth HJ, Borst UE, Blochinger G, Schmid W (1982) Epidemiology of hepatitis A and B infection in Western European countries and in Germans travelling abroad. In: Szmuness W, Alter HJ, Maynard JE, (eds). Viral Hepatitis 1981 International Symposium. The Franklin Institute Press, Philadelphia, pp 157–167
8. Khuroo MS, Rustgi VK, Dawson GJ, Mushahwar IK, Yattoo GN, Kamili S, Khan BA (1994) Spectrum of hepatitis E virus infection in India. J Med Virol 43: 281–286
9. Lavanchy D, Morel B, Frei PC (1994) Seroprevalence of hepatitis E virus in Switzerland. Lancet 344: 747–748
10. Lemon SM (1992) HAV: current concepts of the molecular virology, immunobiology and approaches to vaccine development. Rev Med Virol 2: 73–87
11. Mahoney FJ, Farley TA, Kelso KY, Wilson SA, Horan M, Mc Farland LM (1992) An outbreak of hepatitis A associated with swimming in a public pool. J Infect Dis 165: 615–618
12. Mc Caustland KA, Bond WW, Bradley DW, Ebert JW, Maynard JE (1982) Survival of hepatitis A virus in faeces after drying and storage for 1 month. J Clin Microbiol 16: 957–958
13. Paul DA, Knigge MF, Ritter A, Gutierrez R, Pilot-Matias T, Chau KH, Dawson GJ (1994) Determination of hepatitis E virus seroprevalence by using recombinant fusion proteins and synthetic peptides. J Infect Dis 169: 801–806
14. Reyes GR, Huang CC, Tam AW, Purdy MA (1993) Molecular organization and replication of hepatitis E virus. In: Kaaden OR, Eichhorn W, Czerny CP (eds). Unconventional agents and unidentified viruses. Recent advances in biology and epidemiology. Springer, Wien New York, pp 15–25 (Arch Virol [Suppl] 7)
15. Robertson BH, Jansen RW, Khanna B, Totsuka A, Nainan OV, Siegl G, Widell A, Margolis HS, Isomura S, Ito K, Ishizu T, Moritsugu Y, Lemon SM (1992) Genetic relatedness of hepatitis A virus strains recovered from different geographic regions. J Gen Virol 73: 1365–1377

16. Shapiro CN, Margolis HS (1993) Worldwide epidemiology of hepatitis A virus infection. J Hepatol 18 [Suppl 2]: 11–14

17. Skaug K, Hagen IJ, von der Lippe B (1994) Three cases of acute hepatitis E virus infection imported into Norway. Scand J Infect Dis 26: 137–139

18. Steffen R, Kane MA, Shapiro CN, Billo N, Schoellhorn KJ, van Damme P (1994) Epidemiology and prevention of hepatitis A in travellers. J Am Med Assoc 272: 885–889

19. Tassopoulos NC, Krawczynski K, Hatzakis A, Katsoulidou A, Delladetsima I, Koutelou MG, Trichopoulos D (1994) Role of hepatitis E virus in the etiology of community-acquired non-A, non-B hepatitis in Greece [case report]. J Med Virol 42: 124–128

20. Tawfik el Zimaity DM, Hyams KC, Imam IZE, Watts DM, Bassily S, Naffea EK, Sultan Y, Emara K, Burans J, Purdy MA, Bradley DW, Carl M (1993) Acute sporadic hepatitis E in an Egyptian pediatric population. Am J Trop Med Hyg 48: 372–376

21. Weitz M, Siegl G (1985) Variation among hepatitis A virus strains. I. Genomic variation detected by T1 oligonucleotide mapping. Virus Res 4: 53–67

22. Winokur PL, Stapleton JT (1992) Immunoglobulin prophylaxis for hepatitis A. Clin Infect Dis 14: 580–588

23. Zaaijer HL, Yin MF, Lelib PN (1992) Seroprevalence of hepatitis E in the Netherlands. Lancet 340: 681

24. Zaaijer HL, Kok M, Lelie PN, Timmermann RJ, Chau K, Vanderpal HJH (1993) Hepatitis E in the Netherlands – imported and endemic. Lancet 341: 826

25. Zanetti AR, Dawson GJ (1994) Hepatitis type E in Italy: a seroepidemiological survey. J Med Virol 42: 318–320

Authors' address: Dr. B. C. A. Langer, Max von Pettenkofer Institute for Hygiene and Medical Microbiology, Ludwig Maximilians University, Pettenkofer Strasse 9a, D-80336 Munich, Federal Republic of Germany.

Arch Virol (1996) [Suppl] 11: 181–183

# Significance of imported hepatitis B virus infections

**W. Jilg**

Institute for Medical Microbiology and Hygiene, University of
Regensburg, Regensburg, Federal Republic of Germany

**Summary.** The risk of imported hepatitis B in Germany and comparable
European countries seems to be low as long as suitable control measures are
taken. These measures include testing for hepatitis B markers of immigrants from
highly endemic areas as well as of individuals who stayed for a longer period of
time in such regions, information of chronic carriers of hepatitis B virus and
vaccination of their contacts, and vaccination of travellers to endemic regions
who run an increased risk for contracting hepatitis B as e.g. i.v drug users or sex
tourists.

## Introduction

On a worldwide scale, hepatitis B is still one of the most important infectious diseases. Nearly 40% of the worlds population experience infection with
hepatitis B virus. More than 300 million individuals are chronic carriers of the
agent and, thus, represent a huge virus reservoir. About 6 millions die every year
due to the sequelae of acute or chronic hepatitis B such as fulminant hepatitis,
chronic liver failure, cirrhosis of the liver or hepatocellular carcinoma.

Hepatitis B is highly endemic in South East Asia as well as in Central and
South Africa where carrier rates of up to 20% can be observed. The countries of
the Near and Middle East, Northern Africa, Eastern Europe and South America
show a medium degree endemicity with 2–5% of virus carriers, whereas in the
industrialized countries of Europe and North America between 0.2 and 1% of
the population are chronically infected with hepatitis B virus. With an increasing
number of immigrants from highly endemic areas, with more and more individuals from industrialized countries working at least for some time in the third
world, and with a steadily increasing number of tourists visiting areas where
hepatitis B is highly endemic, there is a growing possibility of importation of
hepatitis B infections into regions of low prevalence. The origin of these
infections, their contribution to the virus reservoir in countries with low hepatitis
B prevalence as well as the precautions necessary to prevent importation are
discussed.

## Transmission of hepatitis B

Hepatitis B virus is exclusively transmitted parenterally. Vehicles and modes of transmission include blood and blood products, injuries with HBV contaminated sharp instruments or needles, and contact of injured skin or mucous membranes with blood or other body fluids containing HBV. Consequently, main routes of transmission are sexual contacts and infection of the newborn by an infected mother during birth. Furthermore, hepatitis B is a nosocomial infection whenever hygienic conditions are poor as is the case in many developing countries. The main source of the virus is blood of acutely ill or chronically infected individuals. The degree of infectivity of a chronic carrier depends mainly on the concentration of infectious virus, which can be as high as or even exceed $10^6$ to $10^8$ virus particles per ml [1]. Transmission rates in needle stick incidents in which chronic carriers are involved are between 2 and 12% [2–4].

In countries with low hepatitis B endemicity as for example Germany, probably the majority of hepatitis B infections are transmitted through sexual contact and needle sharing among i.v. drug abusers. The role of imported infections is less clear. In principle, however, three groups of individuals are likely to introduce hepatitis B virus into this country: immigrants from highly endemic areas, individuals returning from an extended stay in a country with high hepatitis B prevalence, and travellers coming from such countries with high risk behaviour ("drug" and "sex" tourists) or after medical or paramedical treatment.

## Imported hepatitis B virus infections

Between 10 and 15% of immigrants from South-East Asia or from Central and South-Africa are chronic carriers of hepatitis B virus. It is less known, however, that individuals from east European countries also are frequently chronically infected by hepatitis B virus. Recent reports from the World Health Organisation show 0.5–5% chronic virus carriers among residents of the former USSR, about 5% in Bulgaria, and even up to 11% in Romania [5]. Many individuals of German origin immigrated from these countries to Germany during the last years. Therefore, it is of the utmost importance to test immigrants from the said areas for hepatitis B markers; HBV carriers then should undergo a thorough medical examination and be treated if necessary. Susceptible contacts should be vaccinated.

Individuals who stay in a highly endemic area for months or even years, have a significantly increased risk for contracting hepatitis B. Employees and civil servants of the German Foreign Ministry returning from an extended stay in developing countries showed a prevalence of an anti-HBc of 8.0%, whereas only 5.3% of applicants for the service were anti-HBc positive. Similarly, 11.1% of German development aid workers proved to be anti-HBc positive after their return from abroad compared to a mean prevalence of 5.3% in individuals of similar age in Germany [6]. Of 219 French agricultural community and medical volunteers working in Africa, 23 (10.3%) developed HBV infection during a stay abroad of 18 to 30 months [7]. In a similar study it was shown that 9% (21/234) of expatriate married men working for a British company in Southeast Asia became seropositive within 24 months [8]. Based on these data, a monthly incidence of hepatitis B of 80–420 cases per 100 000 can be estimated for individuals from industrialized areas working in developing countries [9].

Only few data exist concerning the hepatitis B risk of the average traveller. In a prospective study comprising 7887 travellers to various developing countries, Steffen et al. [10] found two cases of hepatitis B among 97 individuals who worked abroad, whereas none of 7317 vacationers had acquired hepatitis B. From this and other studies [9] it can be concluded that the risk for an average tourist to contract hepatitis B even in an area of high endemicity seems to be rather low. However, the risk increases considerably in people with high risk behaviour, such as intravenous drug abuse, sexual contacts with the autochthonous population, or undergoing paramedical or extensive medical treatment. Consequently, drug and sex tourists as well as chronically ill people who might need medical care while on vacation should preferentially be vaccinated against hepatitis B.

## References

1. Tabor E, Purcell RH, London WT, Gerety RJ (1983) Use of and interpretation of results using inocula of hepatitis B virus with known infectivity titers. J Infect Dis 147: 531–534
2. Seef LB, Hoofnagle JH (1979) Immunoprophylaxis of viral hepatitis. Gastroenterology 77: 161–182
3. Grady GF, Lee VA, Prince AM, Gitnick GL, Fawaz KA, Vyas GN, Lecitt MD, Senior JR, Galambos JT, Bynum TE, Singleton JW, Clowdus BF, Akdamar K, Aach RD, Winkelman EI, Schiff GM, Hersh T (1978) Hepatitis B immune globulin for accidental exposures among medical personnel: final report of a multicenter controlled trial. J Infect Dis 138: 625–638
4. Werner B, Grady GF (1992) Accidental hepatitis-B-surface-antigen-positive inoculations. Ann Int Med 97: 367–369
5. World Health Organization (1991) Working group on the control of viral hepatitis in Europe. Munich 22–25 April. ICP/OCO 01610650Y
6. Maass G (1983) Definition der Risikogruppen und Bestimmung von Hepatitis B-Markern vor und nach Impfung. In: Deinhardt F, Spiess H (eds) Impfung gegen Hepatitis B. Die Medizinische Verlagsgesellschaft Marburg, pp 23–29
7. Larouze B, Gaudebout C, Mercier E, Lionsquy G, Dazza MC, Elias M, Gaxotte P, Coulaud JP, Ancelle JP (1987) Infection with hepatitis A and B viruses in French volunteers working in tropical Africa. Am J Epidemiol 126: 31–37
8. Dawson DG, Spivey GH, Korelitz JJ, Schmidt RT (1987) Hepatitis B: risk to expatriates in South East Asia. Br Med J Clin Res 294: 547
9. Steffen R (1990) Risk of hepatitis B for travellers. Vaccine [Suppl] 8: S31–S32
10. Steffen R, Rickenbach M, Wilhelm U, Helminger A, Schär M (1987) Health problems after travel to developing countries. J Infect Dis 156: 84–91

Author's address: Prof. Dr. med. W. Jilg, Institut für Medizinische Mikrobiologie und Hygiene, Universität Regensburg, Franz-Josef-Strauß-Allee 11, D-93042 Regensburg, Federal Republic of Germany.

Arch Virol (1996) [Suppl] 11: 185–193

# Genotypes of hepatitis C virus isolates
# from different parts of the world

**E. Schreier**[1], **M. Roggendorf**[2], **G. Driesel**[1], **M. Höhne**[1], and **S. Viazov**[3]

[1] Robert Koch-Institute, Berlin, [2] Institute of Virology, University of Essen,
Essen, Federal Republic of Germany, [3] Ivanovsky Institute of Virology,
Moscow, Russia

**Summary.** Hepatitis C virus (HCV) causes most cases of posttransfusion non-A, non-B hepatitis. HCV isolates were classified by their genetic relatedness into at least six genotypes and a series of subtypes. Methods for typing included amplification of certain genomic regions using universal or type/subtype specific primers, restriction fragment length polymorphism analysis, differential hybridization, nucleotide sequencing, and serologic genotyping. HCV genotypes and their subtypes coexist in various geographic locations but show different prevalences. The identification of genotypes/subtypes is useful for studies on the molecular epidemiology and pathogenesis of HCV infection.

## Introduction

Hepatitis C virus (HCV) is the major cause of parenterally transmitted non-A, non-B hepatitis [7]. Infections are most common in patients with parenteral exposure to blood and blood products, including haemophiliacs, dialysis patients, and i.v. drug users. A high rate of chronicity leads to cirrhosis in about 20% of patients.

The viral genome is a single-stranded, positive sense RNA of about 9.4 kb encoding one large polyprotein of 3010–3033 amino acids [8, 16]. The polyprotein is cleaved into structural proteins [core, envelope 1 (E1), and envelope 2 (E2)] and several non-structural proteins (NS2–NS5). The genomic organization of HCV (Fig. 1) is similar to flavi- and pestiviruses and represents a new genus of the flaviviridae family. Comparisons of published complete and partial genomic sequences led to the classification of HCV isolates into genotypes and subtypes [5, 6, 27, 30, 32]. Isolates of HCV can be classified into at least six major genotypes (1–6) and a series of more closely related subtypes designated a, b, c, etc. (Fig. 2) [33]. This nomenclature recognizes that HCV genotypes have sequence similarities of less than 72%, whereas similarities between clusters of HCV subtype isolates range from 75%–86%. Individual isolates within these clusters show sequence similarity of >88%. Recent HCV isolates

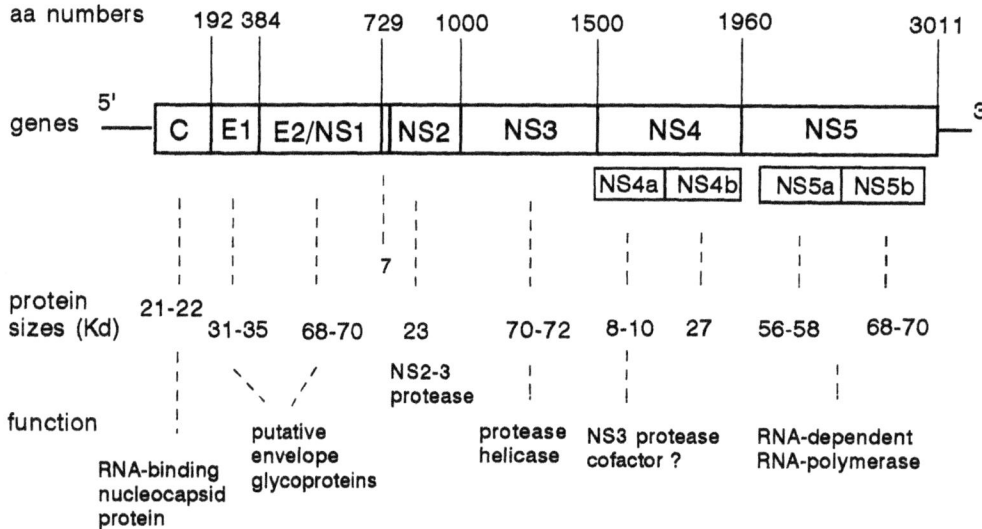

**Fig. 1.** Organization of the hepatitis C viral genome

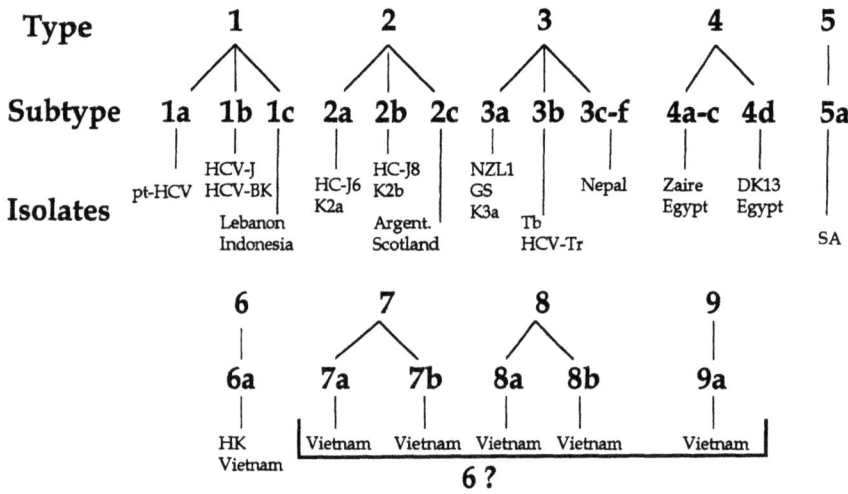

**Fig. 2.** Hepatitis C virus types and subtypes, and their geographical distribution

from commercial blood donors in Vietnam could not be grouped into any of the common genotypes 1 to 6 [36], and these Vietnamese HCV isolates were classified preliminarily as new genotypes 7, 8 and 9.

## HCV genotyping

Various methods of typing HCV have been employed to study isolates from different areas of the world. They include

–the restriction fragment length polymorphism (RFLP) assay [9, 10, 21, 23, 28]
–a reverse-hybridization assay, the line probe assay (LiPA) [35]

–use of universal primers and type/subtype specific oligonucleotide probes (DEIA) [37]

–type-specific oligonucleotide priming [26]

–direct sequencing, with analysis of consensus sequences

–serotyping [31].

Most genotype assays are based on the analysis of polymerase chain reaction (PCR) products. RFLP, LiPA, and DEIA are simple methods and can be used to define quickly the dominant genotype in most cases. We demonstrated that genomic amplification in the 5' noncoding region (5' NCR) followed by selective restriction enzyme digestion with Rsa I and Ava II is useful for differentiating between genotypes 1, 2, and 3 the most prevalent genotypes in Europe [10, 28].

The method of HCV genotyping by using universal primers and type-specific oligonucleotide probes combines, for example, PCR and DNA enzyme immunoassay (DEIA) [37]. In the first step of this method a cDNA of about 250 bp of the HCV core region is amplified by nested PCR. The PCR products are then hybridized to type-specific oligonucleotides fixed to avidin-coated plates from GEN-ETI-K DEIA (Sorin, Biomedica, Saluggia, Italy). The resultant DNA-hybrids are detected by a standard ELISA using monoclonal antibodies reacting with double-stranded DNA [14, 20].

Type-specific oligonucleotide priming as described by Okomoto et al. [26] is simple. However, a significant proportion of isolates may not be typable [18], and in our hands results with at least some of the tested samples were equivocal leading to a significant overestimation of subtype 1b and an underestimation of subtype 1a (A. Widell, pers. comm.).

Direct sequencing and comparison to consensus sequences is unquestionably the most precise method for genotype and subtype analysis but it is not readily applicable to many clinical samples. Sequencing in the NS5 region was especially useful in the study of subtype prevalences in German virus isolates, particularly in intravenous drug abusers (IVDA).

The finding of major antigenic differences between the six genotypes of HCV in the NS4 protein led to developent of a serological method for HCV typing [31], but although such a serotyping assay is an important alternative to conventional PCR-based typing assays the assay presupposes an adequate humoral immune response to HCV.

## Prevalence of HCV genotypes/subtypes

Several HCV types together with their subtypes coexist in various geographic locations but with different prevalences (Fig. 2). Recently, the geographical distribution of the six major types of HCV was investigated in an international collaborative survey [22]. Viral sequences of HCV from blood donors were amplified in the 5'NCR and then typed by RFLP analysis. Donors in European countries (The Netherlands, Scotland, Finland) were almost exclusively infected with type 1, 2, and 3 (Table 1) whereas donors in the Far East (Taiwan, Japan,

**Table 1.** Prevalence of HCV types in different European countries

| Country | No. of samples | No. (%) of samples of HCV types | | | | |
|---|---|---|---|---|---|---|
| Type | | 1 | 2 | 3 | 4 | 5 |
| Scotland | 144 | 67 (47) | 21 (14) | 56 (39) | 0 | 0 |
| Finland | 12 | 3 (25) | 5 (42) | 4 (33) | 0 | 0 |
| The Netherlands | 31 | 18 (58) | 7 (23) | 5 (16) | 1 (3) | 0 |
| Hungary | 47 | 46 (98) | 0 | 0 | 1 (2) | 0 |

McOmish et al. [22]

**Table 2.** Prevalence of HCV types in Far East countries

| Country | No. of samples | No. (%) of samples of HCV types | | | | | |
|---|---|---|---|---|---|---|---|
| Type | | 1 | 2 | 3 | 4 | 5 | 6 |
| Taiwan | 93 | 53 (57) | 40 (43) | 0 | 0 | 0 | 0 |
| Japan | 40 | 31 (77) | 9 (23) | 0 | 0 | 0 | 0 |
| Hong Kong | 37 | 22 (59) | 1 (3) | 0 | 0 | 0 | 12 (38) |
| Macau | 27 | 13 (48) | 4 (15) | 4 (15) | 0 | 0 | 6 (22) |
| Singapore | 10 | 5 (50) | 2 (20) | 3 (30) | 0 | 0 | 0 |
| Thailand | 83 | 35 (42) | 0 | 48 (58) | 0 | 0 | 0 |

McOmish et al. [22]

Hong Kong, Macau, Singapore, and Thailand) were infected mainly with type 1 and in lower frequency with type 2, 3, and 6 (Table 2). The prevalence of certain HCV types seems to be somewhat country-specific: for example, HCV type 4 was found in samples from the Middle East and Egypt, Zaire, and Burundi [4–6, 9, 22, 32, 35], genotype 5 was identified in South Africa [4, 5, 9], and genotype 6 in Hong Kong and Vietnam [5, 9, 22, 36].

The investigations of HCV subtype distribution in six European countries showed a high prevalence of subtype 1b (Table 3). In addition, we found an interesting distribution of HCV subtypes among intravenous drug users (IVDU) [2, 10, 28, 34]. A high percentage of HCV RNA-positive IVDU was infected by HCV subtype 3a which is detected infrequently among non-IVDU HCV-positive individuals (e.g. transfusion recipients) (Table 4). It must be stressed that the predominance of HCV subtype 3a among IVDU was also found in Spain (J. Esteban, pers. comm.), and Sweden (A. Widell, pers. comm.).

It is highly probable that the common use of needles and syringes leads to frequent contact and transmission of HCV within the community of drug users. However, this is only one possible explanation for the high prevalence of subtype 3a among IVDU; as type 3a infection is associated with a higher number of

**Table 3.** HCV subtype distribution in Europe

|  | 1a | 1b | 2a | 2b | 3a | 4a | n.t. |
|---|---|---|---|---|---|---|---|
| Italy<br>n = 19 | 4 (21%) | 9 (47%) | 0 | 0 | 3 (16%) | 2 (12%) | 0 |
| Germany<br>n = 275 | 10 (4%) | 163 (59%) | 10 (4%) | 1 (0.4%) | 74 (28.6%) | 0 | 7 (3%) |
| Spain<br>n = 81 | 8 (10%) | 53 (65%) | 0 | 0 | 12 (15%) | 1 (%) | 7 (9%) |
| Austria<br>n = 45 | 6 (13%) | 32 (71%) | 3 (7%) | 0 | 2 (4%) | 1 (2%) | 1 (2%) |
| Russia<br>n = 37 | 9 (24%) | 26 (70%) | 0 | 0 | 2 (6%) | 0 | 0 |
| Turkey<br>n = 19 | 0 | 17 (90%) | 0 | 0 | 2 (10%) | 0 | 0 |

*n.t.* Not typed

**Table 4.** Prevalence of HCV types in two groups of chronic hepatitis C patients in Germany

| Patients | Types<br>no. | 1 | 2 | 3 |
|---|---|---|---|---|
| IVDU | 136 | 66 (49%) | 8 (6%) | 62 (46%) |
| non-IVDU | 39 | 36 (92%) | 1 (3%) | 2 (6%) |

sexual partners among IVDU, we cannot rule out that sexual transmission contributes to the HCV epidemic among IVDU [34]. IVDU is one of the major reservoirs of HCV infection in Western countries, and the predominance of HCV subtype 3a in this population may have some important epidemiological and clinical implications.

Generally typing HCV isolates may be important in
 –studies of the worldwide molecular epidemiology of HCV
 –tracing HCV infections in risk groups (e.g. i.v. drug users)
 –tracing sources of infections (e.g. in blood products)
 –studies of the relationships between type/subtype and the clinical status and/or outcome of disease
 –clinical studies of the correlation of types/subtypes and pathogenesis of HCV infection
 –study of the significance of types/subtypes in response to antiviral treatment of HCV infection (e.g. interferon).

At this time the spread of hepatitis C virus and the restriction of some genotypes, e.g. types 5 and 6 to certain geographical areas are difficult to understand. More detailed epidemiological studies of HCV infections are needed to gain more insight into a possible type/subtype-specific pathogenesis of HCV.

Although, there is preliminary evidence that genotype 1 is associated with a much poorer response to interferon (IFN) than type 2 and type 3 [3, 38] many other factors may also affect the response to IFN. Sequence divergences in connection with the quasispecies nature of hepatitis C virus genome [11, 19, 25] and titers of circulating HCV RNA [15, 17, 38] may not correlate with infectivity. The quasispecies complexity of the hypervariable region 1 (HVR1) within the putative envelope protein E2 before IFN therapy may be one of the factors affecting the outcome of therapy in chronic hepatitis C patient [1].

Recently, we determined the HCV subtype and also the sequence variations in a group of patients who developed chronic hepatitis C after parenteral administration of a contaminated anti-D immunoglobulin [12]. This immunoglobulin used to protect against the effects of Rh incompatibility in pregnancy infected about 2500 women in East Germany in 1978/79. All isolates investigated were classified as subtype 1b. The contaminated immunoglobulin was infectious, and could be used to infect a human lymphoid cell line (H9), a porcine embryonal kidney cell line (POEK), and human fibroblasts [13, 24, 39].

However, there can be problems in tracing HCV infections by genotyping because a HCV genotype can become selected during transmission or replication in vitro. Recently, we demonstrated genotype selection during transmission of HCV from IVDU mothers to their infants [29]: selection from subtype 3a or 2a dominant in mothers to 1b in infants occurred. Interestingly, replication in vitro in POEK and H9 cell lines inoculated with serum of mothers led to a subtype selection from 3a (2a) to 1b. Although very little is known about the replication and cellular tropism of HCV, types and subtypes of HCV may have different abilities to replicate in cells.

In conclusion, the characterization of hepatitis C virus infections by genotyping has importance in studies of the taxonomy and epidemiology of HCV, the prevention, diagnosis, and therapy of HCV infections, and the development of HCV vaccines.

## References

1. Berg T, König V, Schreier E, Binder T, Hopf U (1994) PCR-SSCP fingerprinting of E2 of HCV in patients treated with interferon-α: a simple method for the assessment of genomic changes. J Hepatol 21: 98
2. Berg T, Hopf U, Stark K, Baumgarten R, Lobeck H, Schreier E (1996) Distribution of hepatitis C virus genotypes in German patients with chronic hepatitis C: correlation with clinical and virological parameters. Hepatology (submitted)
3. Bréchot C, Kremsdorf D (1993) Genetic variation of the hepatitis C virus (HCV) genome: random events or a clinically relevant issue? J Hepatol 17: 265–268
4. Bukh J, Purcell RH, Miller RH (1992) Sequence analysis of the 5′ noncoding region of hepatitis C virus. Proc Natl Acad Sci USA 89: 4942–4946
5. Bukh J, Purcell RH, Miller RH (1993) At least 12 genotypes of hepatitis C virus predicted by sequence analysis of the putative E1 gene of isolates collected worldwide. Proc Natl Acad Sci USA 90: 8234–8238
6. Bukh J, Purcell RH, Miller RH (1994) Sequence analysis of the core gene of 14 hepatitis C virus genotypes. Proc Natl Acad Sci USA 91: 8239–8243

7. Choo QL, Kuo G, Weiner AJ, Overby LR, Bradley DW, Houghton M (1989) Isolation of a cDNA clone derived from a blood-borne non-A, non-B hepatitis genome. Science 244: 359–362

8. Choo QL, Richman KH, Han JH, Berger K, Lee C, Dong C, Gallegos C, Coit D, Medina-Selby A, Barr PJ, Weiner AJ, Bradley DW, Kuo G, Houghton M (1991) Genetic organization and diversity of the hepatitis C virus. Proc Natl Acad Sci USA 88: 2451–2455

9. Davidson F, Simmonds P, Ferguson JC, Jarvis LM, Dow BC, Follett C, Seed CRG, Krusius T, Lin C, Medgyesi GA, Kiyokawa H, Olim G, Duraisamy G, Cuypers T, Saeed AA, Teo D, Conradie J, Kew MC, Lin M, Nuchaprayoon C, Ndimbie OK, Yap PL (1995) Survey of major genotypes and subtypes of hepatitis C virus using RFLP of sequences amplified from the 5' non-coding region. J Gen Virol 76: 1197–1204

10. Driesel G, Wirth D, Stark K, Baumgarten R, Sucker U, Schreier E (1994) Hepatitis C virus (HCV) genotype distribution in German isolates: studies on the sequence variability in the E2 and NS5 region. Arch Virol 139: 379–388

11. Higashi Y, Kakumu S, Yoshioka K, Wakita T, Mizokami M, Ohba K, Ito S, Ishikawa T, Takayanagi M, Nagai Y (1993) Dynamics of genome change in the E2/NS1 region of hepatitis C virus in vivo. Virology 197: 659–668

12. Höhne M, Schreier E, Roggendorf M (1994) Sequence variability in the env-coding region of hepatitis C virus isolated from patients infected during a single source outbreak. Arch Virol 137: 25–34

13. Höhne M, Nissen E, Schreier E (1995) Variability of hepatitis C virus genome in a single source outbreak and during long term replication in cell culture. 4th International conference on current trends in chronically evolving viral hepatitis, Perugia, 30.3.–1.4.1995

14. Imberti L, Cariani E, Bettinardi A, Zonaro A, Albertini A, Primi D (1991) An immunoassay for specific amplified HCV sequences. J Virol Methods 34: 233–243

15. Inokuchi K, Yatsuhashi H, Inoue O, Koga M, Nagataki S, Yano M (1994) Correlation of quantitative HCV-RNA levels using a branched DNA enhanced level amplification assay with therapeutic effects of β-interferon in patients with chronic hepatitis C. Int Hepatol Commun 2: 375–382

16. Kato N, Hijikata M, Ootsuyama Y, Nakagawa M, Ohkoshi S, Sugimura T, Shimotohno K (1990) Molecular cloning of the human hepatitis C virus genome from Japanese patients with non-A, non-B hepatitis. Proc Natl Acad Sci USA 87: 9524–9528

17. Lau J, Davis G, Kniffen J, Qian K-P, Urdea M, Chan C, Mizokami M, Nuewald P, Wilber J (1993) Significance of serum hepatitis C virus RNA levels in chronic hepatitis C. Lancet 341: 1501–1504

18. Lau JYN, Mizokami M, Kolberg JA, Davis GL, Prescott LE, Ohno T, Perrillo RP, Lindsay KL, Gish RG, Qian KP, Kohara M, Simmonds P, Urdea MS (1995) Application of six hepatitis C virus genotyping systems to sera from chronic hepatitis C patients in the United States. J Infect Dis 171: 281–289

19. Lu M, Funsch B, Wiese M, Roggendorf M (1995) Analysis of hepatitis C virus quasispecies populations by temperature gradient gel electrophoresis. J Gen Virol 76: 881–887

20. Mantero G, Zonaro A, Albertini A, Bertolo P, Primi D (1991) DNA enzyme immunoassay: general method for detecting products of polymerase chain reaction. Clin Chem 37: 422–429

21. McOmish F, Chan S-W, Dow BC, Gillon J, Frame WD, Crawford FJ, Yap PL, Follett EAC, Simmonds P (1993) Detection of three types of hepatitis C virus in blood donors: investigation of type-specific differences in serological reactivity and rate of alanine aminotransferase abnormalities. Transfusion 33: 7–13

22. McOmish F, Yap PL, Dow BC, Follett EAC, Seed C, Keller AJ, Cobain TJ, Krusius T, Kolho E, Naukkarinen R, Lin C, Lai C, Leong S, Medgyesi GA, Hejjas M, Kiyokawa H, Fukada K, Cuypers T, Saeed AA, Alrasheed AM, Lin M, Simmonds P (1994) Geographical distribution of hepatitis C virus genotypes in blood donors – an international collaborative survey. J Clin Microbiol 32: 884–892

23. Nakao T, Enomoto N, Takada N, Takada A, Date T (1991) Typing of hepatitis C virus genomes by restriction fragment length polymorphism. J Gen Virol 72: 2105–2112

24. Nissen E, Höhne M, Schreier E (1994) In vitro replication of hepatitis C virus in a human lymphoid cell line (H9). J Hepatol 20: 437

25. Okada S-I, Akahane Y, Suzuki H, Okamoto H, Mishiro S (1992) The degree of variability in the amino terminal region of the E2/NS1 protein of hepatitis C virus correlates with responsiveness to interferon therapy in viremic patients. Hepatology 16: 619–624

26. Okamoto H, Sugiuama Y, Okada S-I, Kurai K, Akahane Y, Sugai Y, Tanaka T, Sato K, Tsuda F, Miyakawa Y, Mayumi M (1992b) Typing hepatitis C virus by polymerase chain reaction with type-specific primers: application to clinical surveys and tracing infectious sources. J Gen Virol 73: 673–679

27. Sakamoto M, Akahane A, Tsuda F, Tanaka T, Woodfield DG, Okamoto H (1994) Entire nucleotide sequence and characterization of hepatitis C virus of genotype V/3a. J Gen Virol 75: 1761–1768

28. Schreier E, Driesel G, Höhne M (1993) Hepatitis-C-Virus Studien zur Genotyp-differenzierung, -charakterisierung und -verteilung. Bundesgesundheitsamt, pp 136–137

29. Schreier E, Höhne M, Nissen E, Pirmann M (1994) HCV genotype selection during vertical transmission and in vitro replication. 2nd International meeting on hepatitis C and Related Viruses. San Diego, 31.7.–5.8.1994, Abstract 285

30. Simmonds P, Holmes EC, Cha TA, Chan SW, McOmish F, Irvine B, Beall E, Yap PL, Kolberg J, Urdea MS (1993) Classification of hepatitis C virus into six major genotypes and a series of subtypes by phylogenetic analysis of the NS-5 region. J Gen Virol 74: 2391–2399

31. Simmonds P, Rose KA, Grahm S, Chan S-W, McOmish F, Dow BC, Follett EAC, Yap PL, Marsden H (1993) Mapping of serotype-specific, immunodominant epitopes in the NS-4 region of hepatitis C virus (HCV): use of type-specific peptides to serologically differentiate infections with HCV type 1, 2 and 3. J Clin Microbiol 31: 1493–1503

32. Simmonds P, McOmish F, Yap PL, Chan SW, Lin CK, Dusheiko G, Saeed AA, Holmes EC (1993) Sequence variability in the 5′ non coding region of hepatitis C virus: identification of a new virus type and restrictions on sequence diversity. J Gen Virol 74: 661–668

33. Simmonds P, Alberti A, Alter HJ, Bonino F, Bradley DW, Brechot C, Brouwer JT, Chan SW, Chayama K, Chen DS, Choo QL, Colombo M, Cuypers HTM, Date T, Dusheiko GM, Esteban JI, Fay O, Hadziyannis SJ, Han J, Hatzakis A, Holmes EC, Hotta H, Houghton M, Irvine B, Kohara M, Kolberg JA, Kuo G, Lau JYN, Lelie PN, Maertens G, McOmish F, Miyamura T, Mizokami M, Nomoto A, Prince AM, Reesink HW, Rice C, Roggendorf M, Schalm SW, Shikata T, Shimotohno K, Stuyver L, Trepo C, Weiner A, Yap PL, Urdea MS (1994) A proposed system for the nomenclature of hepatitis C viral genotypes. Hepatology 19: 1321–1324

34. Stark K, Schreier E, Driesel G, Müller R, Wirth D, Bienzle U (1995) Prevalence and determinants of anti-HCV seropositivity and of HCV genotype among intravenous drug users in Berlin. Scand J Infect Dis 27: 331–337

35. Stuyver L, Rossau R, Wyseur A, Duhamel M, Vanderborght B, Van Heuverswyn H, Maertens G (1993) Typing of hepatitis C virus isolates and characterization of new subtypes using a line probe assay. J Gen Virol 74: 1093–1102

36. Tokita H, Okamoto H, Tsuda F, Song P, Nakata S, Chosa T, Iizuka H, Mishiro S, Miyakawa Y, Mayumi J (1994) Hepatitis C virus variants from Vietnam are classifiable into the seventh, eighth, and ninth major genetic groups. Proc Natl Acad Sci USA 91: 11022–11026

37. Viazov S, Zibert A, Ramakrishnan K, Widell A, Cavicchini A, Schreier E, Roggendorf M (1994) Typing of hepatitis C virus isolates by DNA enzyme immunoassay. J Virol Methods 48: 81–91

38. Yoshioka K, Kakumu S, Wakita T, Wakita T, Ishikawa T, Itoh Y, Takayanagi M, Higashi Y, Shibata M, Morishima T (1992) Detection of hepatitis C virus by polymerase chain reaction and response to interferon-alpha therapy: relationship to genotype virus: J Hepatol 16: 293–299

39. Zibert A, Schreier E, Roggendorf M (1995) Antibodies in human sera specific to hypervariable region 1 of hepatitis C virus can block viral attachment. Virology 208: 653–661

Authors' address: Dr. E. Schreier, Robert Koch-Institute, Nordufer 20, D-13353 Berlin, Federal Republic of Germany.

Arch Virol (1996) [Suppl] 11: 195–202

# HIV-1 subtype O: epidemiology, pathogenesis, diagnosis, and perspectives of the evolution of HIV

**L. G. Gürtler**[1], **L. Zekeng**[2], **J. M. Tsague**[2], **A. von Brunn**[1], **E. Afane Ze**[2], **J. Eberle**[1], and **L. Kaptue**[2]

[1] Max von Pettenkofer Institute for Hygiene and Medical Microbiology, University of Munich, Federal Republic of Germany,
[2] Centre Hospitalier Universitaire, Université 1 de Yaoundé, Cameroon

**Summary.** HIV-1 subtype O is a new HIV variant originating in the West-Central African region, with highest prevalences in countries such as Cameroon, Equatorial Guinea and Gabon. Detection of antibodies to HIV-1 subtype O can pose problems in unmodified ELISA tests, and confirmation of anti-HIV-1 subtype O in immunoblot may give false negative results in some specimens. Nucleic acid-based assays designed for HIV-1 detection do not amplify or detect sequences from HIV-1 subtype O. In their env sequences, HIV-1 subtype O strains show a higher heterogeneity than the classical HIV-1 subtypes, leading to the conclusion that HIV-1 subtype O has been introduced into the human population only recently. Further, unidentified subtypes are also likely to exist.

## Types and subtypes of HIV

Human immunodeficiency viruses (HIV) are divided in two types, HIV-1 and HIV-2. HIV-1 was isolated in 1983, and from the clustering of sequence heterogeneity, especially in the env region, has been subdivided into subtypes A to H. Subtypes A, G and E, and B and D, are more closely related. HIV-1 subtype O was described only recently, and is found in West Central Africa and France. HIV-1 viruses have a genomic arrangement with LTR, gag, pol env and nef and again LTR; in contrast to HIV-2, the nef and env genes do not overlap. HIV-1 subtype O has a genomic arrangement characteristic of HIV-1, including vpu but not vpx as activating factor. HIV-2 is subdivided into subtypes A and B, and there is evidence of further subtypes C, D and E [12]. The 25 isolates of HIV-1 subtype O characterized today show a high degree of heterogeneity, suggesting that further ranking within this subtype will be necessary.

### Nomenclature

HIV-1 subtype O is grouped with the other subtypes A to H in HIV-1. Currently, French investigators call HIV-1 subtype O group O, to distinguish it from the other HIV-1s which are designated HIV-1 group M. M stands for major, whereas O stands for the letter in the alphabet and marks the distance of HIV-1 subtype O to the other HIV-1 subtypes. The letter O was chosen to leave room between H and O, and O to Z, for grouping new viruses.

## Epidemiology

HIV-1 has been spreading, at least in Central Africa, since 1976. Subtype B is distributed worldwide, whereas all other subtypes are found in Central Africa with varying prevalences [7, 8]; most prevalent is subtype A, followed by subtypes B and D. Subtype C and E are more common in the southern parts of Central Africa, and G and H in the western regions [9]. Outside Africa, subtype E is highly prevalent in Thailand and India, subtype C in India, and subtype F in Romania and Brazil. Prevalences of HIV-1 in urban centers in Central Africa may reach over 30% in the sexually active population, whereas in some rural regions HIV infections are still very rare.

HIV-2 is prevalent mostly in West Africa, also in Mozambique and India, and in Europe especially in France and Portugal, i.e. in countries with historical links to African regions where HIV-2 is present. HIV-2 is less pathogenic than HIV-1 and, consequently, the epidemic of HIV-2 has not had the impact of HIV-1. In contrast to HIV-1, HIV-2 is seldom transmitted vertically from mother to child [7].

The first HIV-1 subtype O was isolated from a Cameroonian patient in 1987 by M. Peeters and G. van der Groen in Antwerp [13]. HIV-1 subtype O has its epicenter in Cameroon and the neighbouring countries, such as Equatorial Guinea and Gabon. In Cameroon, HIV-1 subtype O is responsible for 10% of all HIV-1 infections [15]. Two HIV-1 subtype O infections have been found in prostitutes in Niamey, the capital of Niger (E. Delaporte, pers. comm., 1994), and there are also claims that HIV-1 subtype O is present in Ghana. In Europe, 16 HIV-1 subtype O infections were reported from France [6, 11], 5 from Belgium and one from Germany. All European HIV-1 subtype O-infected patients have links to West Central Africa.

If compared to the distribution and rapid spread of other HIV-1 subtypes, HIV-1 subtype O is presently not an epidemiological problem of highest priority, but since some anti-HIV-1 subtype O specimens escape detection by the currently used diagnostic assays, the subtype O may become a problem, not only in West Central Africa, if screening tests are not properly modified.

## Pathogenesis

The pathogenic potential of HIV-1 subtype O to cause immunodeficiency is very similar to that of the other HIV-1 subtypes. Of 15 Cameroonian patients under study for 2 years by our group, five developed AIDS and three died. Symptoms were typical and included tuberculosis, diarrhoea, weight loss, chronic cough and lymphoma; Kaposi sarcoma has also been observed. Vertical HIV-1 subtype O transmission, one from Cameroon and the other from France, has been reported. The first HIV-1 subtype O-infected person identified in France in 1992 died from AIDS [1]. In conclusion, HIV-1 subtype O seems to be as pathogenic as the other HIV-1 subtypes and more pathogenic than HIV-2. Preliminary sequence data also suggest that the mutation rate in HIV-1 subtype O is higher than in HIV-1 subtype B (B. Korber, pers. comm.). The patient from whom the first HIV-1 subtype O was isolated in 1987 is still alive but with progressive depletion of immune function.

### Diagnostic problems of HIV-1 subtype O infection

*Detection of HIV-1 subtype O antibodies*

#### Indirect and double antigen ELISA

Most of the currently commercially available, indirect HIV-1 + 2 ELISA detect HIV-1 subtype O antibodies without problems, although the extinctions read with such tests can be close to the cut-off. Nevertheless, some assays yield false negative results with some specimens. When ELISA tests based on the double antigen principle (sandwich Elisa or so-called third generation ELISA) were used, about 20% of the anti-HIV-1 subtype O-positive specimens gave false negative results [6], but assays have been changed or are in the process of being changed. Since 1995, commercial assays have been available in which HIV-1 subtype O antigen is included [10, 11]. Double antigen ELISA are preferable to indirect ELISA tests because they detect simultaneously low titered IgG and IgM antibodies, and so can identify developing serconversion about one week earlier.

#### Confirmatory tests

Problems arise when confirmatory tests are needed for HIV-1 subtype O antibodies because neither specific tests for detecting anti-HIV-1 subtype O by indirect immunofluorescence nor immunoblot tests are currently commercially available.

*Immunofluorescence*: For HIV-1 infected cells, coated on microscopic slides, complete epitope sets for the attachment of anti-HIV-1 subtype O antibodies are available, so false negative results are not obtained but in some specimens the reaction is very faint [3]. Use of cells infected additionally with HIV-1 subtype O could circumvent this problem.

*Immunoblot*: Dependent on the crossreactivity of HIV-1 subtype O antibodies, all variations from a full banding pattern to only weakly stained bands can be observed on immunoblot strips coated with HIV-1 subtype B antigens. Due to conserved structures, staining is most prominent with proteins of the integrase (p 34) and reverse transcriptase (p51, p66). Within the HIV-1 glycoproteins, reaction is best with gp160. In some specimens, crossreactivity with the HIV-1 proteins is so faint that the immunoblot result is negative. Reactivity of anti-HIV-1 subtype O with HIV-2 proteins on immunoblot strips is very low; if at all,

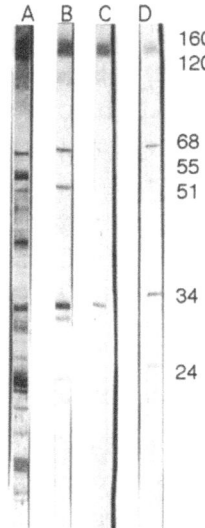

**Fig. 1.** Typical profile of anti-HIV-1 subtype O sera following reaction with a HIV-1 subtype B immunoblot. *A* anti-HIV-1 subtype B, positive-control; *B* anti-HIV-1 subtype O specimen with a full banding pattern; *C* anti-HIV-1 subtype O specimen with a reduced staining of the bands; *D* anti-HIV-1 subtype O specimen revealing a weak profile that, by usual criteria, would hardly be considered positive. Numbers on the right side of the strips indicate the molecular weights of the glycoproteins (gp 160, gp 120), the reverse transcriptase (p68, p51) and the integrase (p34) in thousands

reactivity is observed with gp130 and gp36 [1, 4, 6]. The reaction profile of anti-HIV-1 subtype O with HIV-1 subtype B antigens on immunoblot strips is shown in Fig. 1.

Changes in the antigen composition of immunblots are urgently needed because until now, no manufacturer has added HIV-1 subtype O antigens to the immunoblot. Such specific HIV-1 subtype O immunoblots are presently only available in some research laboratories.

### Detection of virus

### Proteins

HIV-1 subtype O p24 antigen is detected when using a HIV-1 p24 antigen assay; monoclonal antibodies of the commercially available assays show about 50% of the reactivity found with HIV-1 subtype B, so sufficient reactivity is available to detect more than 20 pg/mL antigen, equivalent to $2 \times 10^5$ virus particles [1, 3]. A further method to detect HIV-1 subtype O is determination of the reverse transcriptase: since the enzyme uses the same magnesium concentration to transcribe the RNA template, this assay has the same sensitivity for all HIV-1 subtypes.

Nucleic acids

Respective procedures detect HIV by its nucleic acid in the viral particle (RNA) or in the infected cell (DNA). Assays for HIV-1 nucleic acid detection are polymerase chain reaction (PCR), branched DNA signal amplification assay (DNA SAA) and isothermic nucleic acid amplification assay (NASBA). None of these assays presently detects HIV-1 subtype O RNA or DNA since the divergence of the HIV-1 subtype O sequences is so great that primers selected for the amplification of the other HIV-1 subtype genomes do not hybridize. Primers that amplify HIV-1 subtype O DNA properly have been designed for the gag and env regions [11].

*Conclusion based on the detection of HIV-1 subtype O*

According to presently available data, HIV-1 subtype O is spreading in West Central Africa at least since the early eighties, and the question arises why HIV-1 is prevalent predominantly in Central Africa, HIV-2 in West Africa, and HIV-1 subtype O in West Central Africa. One hypothesis is that HIVs are old viruses, selected by evolution in different primate species for thousands of years, and transmitted to humans only recently. Intraprimate transmission in monkeys occurs by hunting and eating young animals especially by other species and/or by contact of open lesions with blood [12]. A speculation for the long existence of immunodeficiency-like viruses is drawn from the fact that besides chimpanzees [5] several mammalian species carry immunodeficiency viruses or closely related viruses such as the visna/maedi virus of sheep, the equine infectious anaemia virus of horse and the bovine immunodeficiency virus of cattle. When a vector allowing transspecies transmission is excluded, an evolution of those viruses in parallel with their host over more than 20 million years in primates might have been possible [14] (Fig. 2).

An argument with some validity for this hypothesis is the high HIV-2 prevalence in some areas in West Africa, where the most closely related virus to HIV-2, the SIV-2, is endemic in monkeys like the sooty mangabey monkeys, African green monkeys, mandrill and baboon. In mangabey monkeys, the SIV-2 prevalence may reach 15% and intraspecies transmission is most likely by sexual contact, vertical transmission and blood contact during fighting [12]. In Central Africa, only chimpanzees have been found naturally infected with HIV-1 (SIV-1); 1 in 50 animals seems to be infected without signs of immunodeficiency [5, 13]. HIV-1 subtype O is positioned in the evolutionary tree between both chimpanzee viruses, the CPZgab (from Gabon) and the CPZant (from Zaire) (M. Peeters, pers. comm.). Naturally SIV-infected monkeys usually do not develop immunodeficiency, although they might exhibit disease when they are infected with a SIV from another species, e.g. infection of macaque monkeys with the virus from sooty mangabey monkeys [2]. Low pathogenicity in the SIV's natural host argues for a virus with a long history of evolution.

Separation of American monkeys from the Old World monkeys happened about 25 million years ago, and South American monkeys are not infected with

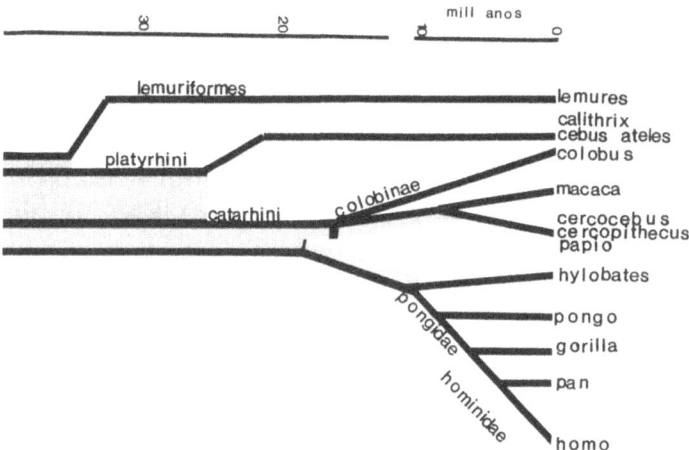

**Fig. 2.** Evolutionary tree of primates. The time scale is in million years, providing a crude scale of the evolutionary age of the different species. The name of the species is given with the Latin name: Cercocebus is the sooty magabey monkey, Cercopithecus is the African green monkey, Papio is the species of baboon and mandrill; Hylobates is the gibbon, Pongo the orang and Pan the chimpanzee. The group of the cercopithecinae with Cercocebus, Cercopithecus and Papio is naturally infected with an immunodeficiency virus very closely related to HIV-2 (proposed name SIV-2), whereas the chimpanzee is rarely infected with a HIV-1 like virus (tentatively named SIV-1). The South American Platyrhimi monkeys are free of HIV, as are monkeys living in Asia (Hylobates, Pongo and Macaca) and Madagascar (Lemures)

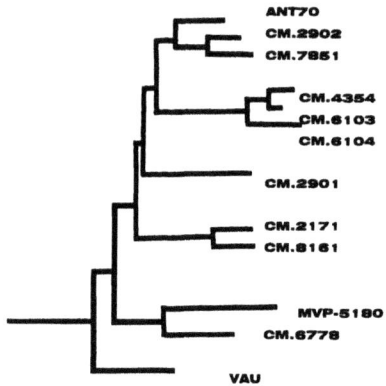

**Fig. 3.** Evolutionary tree of some HIV-1 subtype O isolates from Cameroonian patients in 1994, in comparison with the positions of the published strains ANT70, MVP-5180 and VAU. The C2/V3 region of the env gene was amplified by PCR, sequenced and aligned as described [8]. Alignment data of this tree are based mainly on work of E. Allen and B. Korber, Los Alamos (publication pending). CM4354, CM6103 and CM6104 are three consecutive isolates from the same patient within three months, indicating the extreme heterogeneity of coexisting quasispecies. Since only part of the genome has been aligned, the branching of the tree is still preliminary

a HIV-like virus but with HTLV-2. About 10 million years ago, monkeys such as the colobus and cercopithecus species diverged from the common tree, and in the cercopithecinae the above mentioned African species naturally carry SIV-2, and additionally STLV-1. The macaque monkeys also belong to the cercopithecus group but live in Asia, and Asian monkeys are not infected with a HIV-like virus, but with STLV-1.

Compared to the HIV-1 subtypes A to H, the heterogeneity of V3-loop sequences of HIV-1 subtype O is even greater. This diversity of the principal neutralization domain of HIV-1 subtype O hints to a nonequilibrium state between the pathogenic action of the virus and the defense mechanism of its human host. An example of sequence heterogeneity is given from the consecutive bleedings of one Cameroonian patient, isolates CM4354, CM6103 and CM6104 (Fig. 3). This heterogeneity and the high pathogenicity of this HIV-1 subtype O support the hypothesis of recent introduction of this virus to the human species.

## Acknowledgements

It is pleasure to thank those who collaborated in collecting data about HIV-1 subtype O. In Paris: F. Simon, S. Saragosti, M. Chaix and F. Brun-Vézinet; in Antwerp: G. van der Groen, M. vanden Haesevelde and J. Nkengasong; in Montpellier: E. Delaporte and M. Peeters and in Los Alamos: E. Allen, B. Korber and G. Myers.

## References

1. Charneau P, Borman AM, Quillent C, Guétard D, Chamaret S, Cohen J, Rémy G, Montagnier L, Clavel F (1994) Isolation and envelope sequence of a highly divergent HIV-1 isolate: definition of a new HIV-1 group. Virology 205: 247–253

2. Dewhurst S, Embretson JE, Anderson DC, Mullins Ji, Fultz PN (1990) Sequence analysis and acute pathogenicity of molecularly cloned SIV smm-pbj 14. Nature 345: 636–640

3. Gürtler LG, Hauser PH, Eberle J, von Brunn A, Knapp S, Zekeng L, Tsague JM, Kaptue L (1994) A new subtype of human immunodeficiency virus type 1 (MVP-5180) from Cameroon. J Virol 68: 1581–1585

4. Gürtler LG, Zekeng L, Simon F, Eberle J, Tsague JM, Kapture L, Brust S, Knapp S (1995) Reactivity of five anti-HIV-1 subtype O specimens with six different anti-HIV screening ELISAs and three immunoblots. J Virol Methods 51: 177–184

5. Janssens W, Fransen K, Peeters M, Heyndrickx L, Motte J, Bedjabaga L, Delaporte E, Piot P, van der Groen G (1994) Phylogenetic analysis of a new chimpanzee lentivirus SIV cpz-gab2 from a wild captured chimpanzee from Gabon. AIDS Res Hum Retroviruses 10: 1191–1192

6. Loussert-Ajaka I, Ly TD, Chaix ML, Ingrand D, Saragosti S, Courouce AM, Brun-Vézinet F, Simon F (1994) HIV-1/HIV-2 seronegativity in HIV-1 subtype O infected patients. Lancet 343: 1393–1394

7. Marlink R (1994) The biology and epidemiology of HIV-2. In: Essex M, Mboup S, Kanki PJ, Kalengayi MR (eds) AIDS in Africa. Raven Press, New York, pp 47–65

8. Myers G, Korber B, Wain-Hobson S, Jeang KT, Henderson LE, Pavlakis GN (1994) Human Retroviruses and AIDS. A compilation and analysis of nucleic acid and aminoacid sequences. Los Alamos National Laboratory, New Mexico

9. Nkengasong J, Janssens W, Heyndirckx L, Fransen K, Ndumbe P, Motte J, Leonaers A, Ngolle M, Ayuk J, Piot P, van der Groen G (1994) Genotypic subtypes of HIV-1 in Cameroon. AIDS 8: 1405–1412

10. Schable C, Zekeng L, Pau CP, Hu D, Kaptue L, Gurtler L, Dondero T, Tsague JM, Schochetman G, Jaffe H, George JR (1994) Sensitivity of United States HIV antibody tests for detection of HIV-1 group O infections. Lancet 344: 1333–1334

11. Simon F, Ly TD, Baillou-Beaufils A, Fauveau V, Saint-Martin JD, Loussert-Ajaka I, Chaix ML, Saragosti S, Courouce AM, Ingrand D, Janot C, Brun-Vézinet F (1994) Sensitivity of screening kits for anti-HIV-1 subtype O antibodies. AIDS 8: 1628–1629

12. Sharp PM, Robertson DL, Gao F, Hahn BH (1994) Origins and diversity of human immunodeficiency viruses. AIDS [Suppl 1] 8: S27–S42

13. Vanden Haesevelde M, Decourt JL, De Leys J, Vanderborght B, van der Groen G, van Heuverswijn H, Saman E (1994) Genomic cloning and complete sequence analysis of a highly divergent African human immunodeficiency virus isolate. J Virol 68: 1586–1596

14. Wood B (1994) The oldest hominid yet. Nature 371: 280–281

15. Zekeng L, Gürtler L, Afane Ze E, Sam-Abbenyi A, Mbouni-Essomba G, Mpoudi-Ngolle E, Monny-Lobe M, Tapko JB, Kaptue L (1994) Prevalence of HIV-1 subtype O infection in Cameroon: preliminary results. AIDS 8: 1626–1628

Authors' address: Dr. L. Gürtler, Max von Pettenkofer Institut für Hygiene und Medizinische Mikrobiologie, Universität München, Pettenkofer Strasse 9A, D-80336 München, Federal Republic of Germany.

# SpringerVirology

O.-R. Kaaden, W. Eichhorn,
C.-P. Czerny (eds.)

## Unconventional Agents and Unclassified Viruses
Recent Advances in Biology and Epidemiology

1993. 79 partly coloured figures.
VIII, 308 pages. ISBN 3-211-82480-4
Soft cover DM 260,–, approx. US $ 163.00*
Archives of Virology / Supplement 7

The contributions to this volume focused on prion-related diseases, with special emphasis on bovine spongiform encephalopathy and human spongiform encephalopathies, and Borna disease virus, an agent known since long time to be pathogenic for horses and sheep, which is now discussed as a potential pathogen for humans.

P. P. Liberski

## The Enigma of Slow Viruses
Facts and Artefacts

1993. 56 figures. XVI, 277 pages.
ISBN 3-211-82427-8
Soft cover DM 250,–, approx. US $ 179.00*
Archives of Virology / Supplement 6

This comprehensive review covers all aspects of slow unconventional virus infections known today. It includes numerous historical data, biochemistry and molecular biology of the prion protein and its gene, the role of genetics and mutations within PrP gene, spreading and targeting of the virus, biochemistry and neurochemistry of the alterations of different neurotransmitter system and neuropathology.

O. W. Barnett (ed.)

## Potyvirus Taxonomy

1992. 57 figures. IX, 450 pages.
ISBN 3-211-82353-0
Soft cover DM 290,–, approx. US $ 193.00*
Archives of Virology / Supplement 5

This book brings together the collaborative efforts of experts summarizing characteristics of potyviruses which relate to their taxonomy and pointing out areas which require consideration before an international consensus can be reached to help develop effective control strategies against these viruses.

C. De Bac, W. H. Gerlich,
G. Taliani (eds.)

## Chronically Evolving Viral Hepatitis

1992. 72 figures. XIV, 348 pages.
ISBN 3-211-82350-6
Soft cover DM 260,–, approx. US $ 173.00*
Archives of Virology / Supplement 4

Specialists from different medical fields provide a synopsis of all important aspects of viral hepatitis and discuss the clinical significance of newly developed serological assays for diagnosis and prevention.

B. Liess, V. Moennig,
J. Pohlenz, G. Trautwein (eds.)

## Ruminant Pestivirus Infections
Virology, Pathogenesis, and Perspectives of Prophylaxis

1991. 78 figures. VIII, 271 pages.
ISBN 3-211-82279-8
Soft cover DM 220,–, approx. US $ 152.00*
Archives of Virology / Supplement 3

Clinicians, epidemiologists, pathologists, virologists, and molecular biologists present exciting developments in pestivirology as well as perspectives for effective control strategies.

C. H. Calisher (ed.)

## Hemorrhagic Fever with Renal Syndrome, Tick- and Mosquito-Borne Viruses

1991. 75 figures. VII, 347 pages.
ISBN 3-211-82217-8
Soft cover DM 258,–, approx. US $ 166.00*
Archives of Virology / Supplement 1

In this truly international collection information is provided on the molecular biology, antigenicity, diagnosis, epidemiology, clinical aspects, pathogenesis, vaccines, and other aspects of arbovirology.

*\* 10 % price reduction for subscribers to the journal "Archives of Virology"*

SpringerWienNewYork

P.O.Box 89, A-1201 Wien • New York, NY 10010, 175 Fifth Avenue
Heidelberger Platz 3, D-14197 Berlin • Tokyo 113, 3-13, Hongo 3-chome, Bunkyo-ku

# SpringerVirology

F.A. Murphy, C.M. Fauquet,
D.H.L. Bishop, S.A. Ghabrial,
A.W. Jarvis, G.P. Martelli,
M.A. Mayo, M.D. Summers (eds.)

M.A. Brinton, C.H. Calisher,
R. Rueckert (eds.)

W. H. Gerlich (ed.)

## Virus Taxonomy

Classification and
Nomenclature of Viruses
Sixth Report of the International
Committee on Taxonomy of Viruses

## Positive-Strand
## RNA Viruses

## Research in
## Chronic Viral Hepatitis

1995. 185 figures. IX, 586 pages.
Soft cover DM 160,–, approx. US $ 98.00*
ISBN 3-211-82594-0
Archives of Virology / Supplement 10

1994. 182 figures. X, 558 pages.
Soft cover DM 380,–, approx. US $ 224.00*
ISBN 3-211-82522-3
Archives of Virology / Supplementum 9

1993. 46 partly coloured figures. XI, 304 pages.
Soft cover DM 250,–, approx. US $ 156.00*
ISBN 3-211-82497-9
Archives of Virology / Supplementum 8

The Committee's Sixth Report includes one order, 71 families, 11 subfamilies, and 175 genera and more than 4,000 member viruses. On 600 printed pages large amounts of molecular biologic data, illustrated by micrographs and virion diagrams, gene maps and tables give a comprehensive overview and prove helpful in teaching, in diagnostics, in scholarly research, and in the practical areas of medicine, veterinary medicine, plant pathology, insect pest management, and biotechnology.

Positive-strand RNA viruses include the majority of the plant viruses, a number of insect viruses, and animal viruses, such as coronaviruses, togaviruses, flaviviruses, poliovirus, hepatitis C, and rhinoviruses. Works from more than 50 leading laboratories represent latest research on strategies for the control of virus diseases molecular aspects of pathogenesis and virulence; genome replication and transcription; RNA recombination; RNA-protein interactions and host-virus interactions; protein expression and virion maturation; RNA replication; virus receptors; and virus structure and assembly. Highlights include analysis of the picorna-virus IRES element, evidence for long term persistence of viral RNA in host cells, acquisition of new genes from the host and other viruses via copy-choice recombination, identification of molecular targets and use of structural and molecular biological studies for development of novel antiviral agents.

This is an update of the molecular biology and clinical experience on the viruses which cause chronic hepatitis and liver carcinoma in humans and in model animals.

Treatment of chronic hepatitis, reinfection after liver transplantation, in vitro replication of hepatitis B, C and D viruses, immunopathogenesis, variants of hepatitis viruses, oncogenicity, epidemiology and diagnosis and prevention are the major topics of the book.

Thus, clinicians, laboratory physicians, molecular virologists and public health specialists may equally well profit from this book.

*\* 10 % price reduction for subscribers to the journal "Archives of Virology"*

 SpringerWienNewYork

P.O.Box 89, A-1201 Wien • New York, NY 10010, 175 Fifth Avenue
Heidelberger Platz 3, D-14197 Berlin • Tokyo 113, 3-13, Hongo 3-chome, Bunkyo-ku

GPSR Compliance

*The European Union's (EU) General Product Safety Regulation (GPSR)*
*is a set of rules that requires consumer products to be  safe and our*
*obligations to ensure this.*

*If you have any concerns about our products, you can contact us on*
*ProductSafety@springernature.com*

In case Publisher is established outside the EU, the EU authorized
representative is:

Springer Nature Customer Service Center GmbH
Europaplatz 3
69115 Heidelberg, Germany

**Batch number: 09625567**

Printed by Printforce, the Netherlands